Hands-On Standards®, Common Core Edition

Grade 7

Hands-On Standards®, Common Core Edition
Grade 7

hand2mind 78871

ISBN 978-0-7406-9445-5

Vernon Hills, IL 60061-1862

800-445-5985

www.hand2mind.com

Printed in the United States of America.

12 13 14 15 16 17 18 19 20 21 10 9 8 7 6 5 4 3 2

Contents

Statistics and Probability 106

Blackline Masters

Introduction

How do we help students find meaning in mathematics? That is, how do we give students more than a rote script for reciting facts and churning out computations? How do we help students develop understanding?

Hands-On Standards®, Common Core Edition Grade 7 is an easy-to-use reference manual for teachers who want to help students discover meaning in mathematics. Each of the manual's 32 lessons demonstrates a hands-on exploration using manipulatives. The goal is to help students get a physical sense of a problem—to help students get their hands on the concepts they need to know and to help them "see" the meaning.

Each lesson in ***Hands-On Standards*** targets a clearly stated objective. The main part of a lesson offers a story problem that students can relate to and has the students work on the problem using a hands-on approach. Full-color photographs demonstrate the suggested steps. In addition to the main activity, each lesson includes suggested points of discussion, ideas for more exploration, a formative assessment item, and practice pages to help students solidify their understanding. The instructional model is a progression from concrete to abstract.

This book is divided into five sections—Ratios and Proportional Relationships, The Number System, Expressions and Equations, Geometry, and Statistics and Probability. These correspond to the five content domains for Grade 7 as cited in the ***Common Core State Standards for Mathematics***.

Each lesson in this book features one of the following manipulatives:
Algeblocks® • **AngLegs®** • **Centimeter Cubes** • **Color Tiles** • **Deluxe Rainbow Fraction® Circles** • **Deluxe Rainbow Fraction Squares** • **Number Cubes** • **Pattern Blocks** • **Polyhedral Dice** • **Rainbow Fraction Circle Rings** • **Relational GeoSolids®** • **Spinners** • **Two-Color Counters** • **XY Coordinate Pegboard**

Read on to find out how ***Hands-On Standards, Common Core Edition Grade 7*** can help the students in your class find meaning in math and build a foundation for future math success!

A Walk Through a Lesson

Each lesson in ***Hands-On Standards*** includes many features, including background information, objectives, pacing and grouping suggestions, discussion questions, and ideas for further activities, all in addition to the step-by-step, hands-on activity instruction. Take a walk through a lesson to see an explanation of each feature.

Lesson Introduction

A brief introduction explores the background of the concepts and skills covered in each lesson. It shows how they fit into the larger context of students' mathematical development.

Try It! Arrow

In order to provide a transition from the introduction to the activity, an arrow draws attention to the Try It! activity on the next page. When the activity has been completed, return to the first page to complete the lesson.

Objective

The **Objective** summarizes the skill or concept students will learn through the hands-on lesson.

Common Core State Standards

Each lesson has been created to align with one or more of the **Common Core State Standards for Mathematics**.

Talk About It

The **Talk About It** section provides post-activity discussion topics and questions. Discussion reinforces activity concepts and provides the opportunity to make sure students have learned and understood the concepts and skills.

Solve It

Solve It gives students a chance to show what they've learned. Students are asked to return to and solve the original word problem. They might summarize the lesson concept through drawing or writing, or extend the skill through a new variation on the problem.

More Ideas

More Ideas provides additional activities and suggestions for teaching about the lesson concept using a variety of manipulatives. These ideas might be suggestions for additional practice with the skill or an extension of the lesson.

Formative Assessment

Formative assessments allow for on-going feedback on students' understanding of the concept.

LESSON 2

Ratios and Proportional Relationships

Proportional Relationships II

A proportion is an equation that sets two ratios equal to each other. If that equation is true, then the relationship is proportional. Students have checked to see if relationships are proportional by graphing, but now they will check by using their skills with equivalent fractions.

Objective

Determine whether a relationship is proportional by checking for equivalent ratios.

Common Core State Standards

- 7.RP.2a Decide whether two quantities are in a proportional relationship, e.g., by testing for equivalent ratios in a table or graphing on a coordinate plane and observing whether the graph is a straight line through the origin.

Try It! *Perform the Try It! activity on the next page.*

Talk About It

Discuss the Try It! activity.

- **Ask:** *What was the first ratio, or fraction, you built?* (2:8, or $\frac{2}{8}$) *What does the ratio represent?* (the number of green apples to the total number of apples) *What was the second ratio, or fraction, you built?* (1:4, or $\frac{1}{4}$) *What does the ratio represent?* (the number of green apples to the total number of apples)
- **Ask:** *How can you tell if the two ratios are equivalent?* (you can build the fractions and compare or simplify)
- **Ask:** *Is this a proportional relationship? Why or why not?* (yes; the ratios are equivalent)

Solve It

Reread the problem with the students. Have students build the two ratios and draw them on the Fraction Squares BLM. Ask students to explain whether the ratios are equivalent and write an equation to represent the equivalent ratios. (2:8 = 1:4, or $\frac{2}{8} = \frac{1}{4}$)

More Ideas

For other ways to teach about proportional relationships and equivalent ratios—

- Have students use Fraction Tower® Equivalency Cubes to build each ratio. Then, they can compare the heights of the towers to see if they are equivalent and therefore represent a proportional relationship.
- Have students make the fractions using Deluxe Rainbow Fraction® Circles. They can measure the fractions with Rainbow Fraction Circle Rings or compare by stacking to determine whether the fractions are equivalent and therefore represent a proportional relationship.

Formative Assessment

Have students try the following problem.

In Rob's group, there are 2 boys and 3 girls. In Caren's group, there are 4 boys and 6 girls. Which equation shows that the groups are proportional?

A. $\frac{2}{6} = \frac{1}{3}$ **B.** $\frac{2}{5} = \frac{4}{10}$ **C.** $\frac{2}{4} = \frac{3}{6}$ **D.** $\frac{2}{3} = \frac{4}{6}$

12

Try It!

The **Try It!** activity opens with **Pacing** and **Grouping** guides. The **Pacing** guide indicates about how much time it will take for students to complete the activity, including the post-activity discussion. The **Grouping** guide recommends whether students should work independently, in pairs, or in small groups.

Next, the **Try It!** activity is introduced with a real-world story problem. Students will "solve" the problem by performing the hands-on activity. The word problem provides a context for the hands-on work and the lesson skill.

The **Materials** box lists the type and quantity of materials that students will use to complete the activity, including manipulatives such as Color Tiles and Pattern Blocks.

This section of the page also includes any instruction that students may benefit from before starting the activity, such as a review of foundational mathematical concepts or an introduction to new ones.

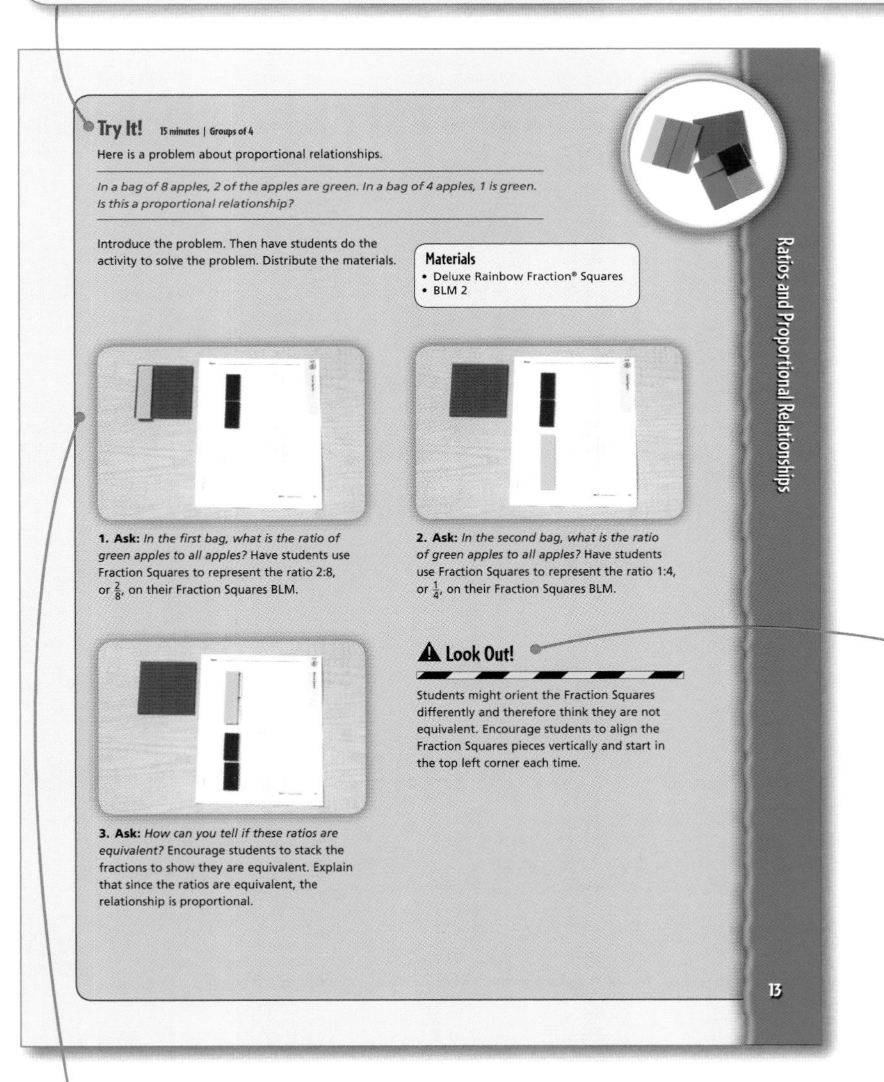

Try It! 15 minutes | Groups of 4

Here is a problem about proportional relationships.

In a bag of 8 apples, 2 of the apples are green. In a bag of 4 apples, 1 is green. Is this a proportional relationship?

Introduce the problem. Then have students do the activity to solve the problem. Distribute the materials.

Materials
- Deluxe Rainbow Fraction® Squares
- BLM 2

1. Ask: *In the first bag, what is the ratio of green apples to all apples?* Have students use Fraction Squares to represent the ratio 2:8, or $\frac{2}{8}$, on their Fraction Squares BLM.

2. Ask: *In the second bag, what is the ratio of green apples to all apples?* Have students use Fraction Squares to represent the ratio 1:4, or $\frac{1}{4}$, on their Fraction Squares BLM.

3. Ask: *How can you tell if these ratios are equivalent?* Encourage students to stack the fractions to show they are equivalent. Explain that since the ratios are equivalent, the relationship is proportional.

Look Out!

Students might orient the Fraction Squares differently and therefore think they are not equivalent. Encourage students to align the Fraction Squares pieces vertically and start in the top left corner each time.

Ratios and Proportional Relationships

13

Look Out!

Look Out! describes common errors or misconceptions likely to be exhibited by students at this age dealing with each skill or concept and offers troubleshooting suggestions.

Step-by-Step Activity Procedure

The hands-on activity itself is the core of each lesson. It is presented in three—or sometimes four—steps, each of which includes instruction in how students should use manipulatives and other materials to address the introductory word problem and master the lesson's skill or concept. An accompanying photograph illustrates each step.

A Walk Through a Student Page

Each lesson is followed by a corresponding set of student pages. These pages take the student from the concrete to the abstract, completing the instructional cycle. Students begin by using manipulatives, move to creating visual representations, and then complete the cycle by working with abstract mathematical symbols.

Exercise

Concrete and **Representational** exercises (pictorial representations of the featured manipulative) help students bridge conceptual learning to symbolic mathematics.

Standards-Based Math Practice

Abstract exercises provide standards-based math practice to allow students to deepen their understanding of the featured skill.

Lesson 2 Ratios and Proportional Relationships

Answer Key

Use Fraction Squares. Complete the model to answer the question. (Check students' work.)

1. In the teacher's pencil jar, there are 10 pencils, 4 of which do not have an eraser. In Julio's pencil bag, there are 5 pencils, 3 of which do not have an eraser. Is the relationship proportional?

$\frac{1}{10}$ $\frac{1}{10}$ $\frac{1}{10}$ $\frac{1}{10}$

No

Using Fraction Squares, model the problem. Draw the model and use it to answer the question.

2. In a bag of 4 instruments, 2 instruments are shakers. In a box of 8 instruments, 4 are shakers. Is the relationship proportional?

Yes

Use Fraction Squares to determine if the relationship is proportional.

3. In PE, 5 of every 6 girls finished a run in less than 10 minutes. Two of every 3 boys finished in less than 10 minutes. Is the relationship proportional?

No

4. Sal paid \$2 for 4 pounds of grapes. Bo paid \$1 for 2 pounds of grapes. Is the relationship proportional?

Yes

Use equivalent ratios to determine if the relationship is proportional.

5. Roberto can ride his bike 4 miles in 20 minutes. Patricia can ride her bike 10 miles in 50 minutes. Is the relationship proportional?

Yes

6. Pearla answered 4 of the 5 questions right on the quiz. Then, on the test, she answered 15 of the 20 questions right. Is the relationship proportional?

No

Download student pages at hand2mind.com/hosstudent.

Extended Response

Extended Response exercises feature an open-ended constructed response question to help teachers gauge student understanding.

Answer Key

Challenge! Ms. Turny's class ratio of boys to girls is 2:3, and Mr. Straight's class ratio of boys to girls is 8:12. Explain how you know if the data portrays a proportional relationship.

Challenge: If the data forms equivalent ratios, then the relationship is proportional. Since 2:3 = 8:12, the data does portray a proportional relationship.

© ETA hand2mind™

Download student pages at hand2mind.com/hosstudent.

Hands-On Standards, Common Core Edition 15

Answers for the Teacher

Answers are provided for teachers on the included student pages.

Student Pages Download

Download clean copies of the student pages by visiting the URL listed.

Ratios and Proportional Relationships

In seventh grade, students extend their understanding of ratios and develop understanding of proportionality to solve a variety of single- and multi-step ratio and percent problems. A **ratio** is a comparison of two numbers or quantities through division and is typically expressed in simplest fraction form (e.g., the ratio of boys to girls in a class is 10 to 15, which would be expressed as 2 to 3). A **proportion** is an equation setting two ratios equal. Blueprint drawings and the actual structures they represent look alike because they are proportionate to one another (e.g., a scale of $\frac{1}{4}$ inch on the drawing to 1 foot of actual length is expressed as $\frac{1}{4}$ = 1'0"). A **proportionality** is a relationship in which the ratio between two quantities does not vary—the ratio remains constant.

At this grade level, students also compute **unit rates**. A *rate* is a ratio that compares two different kinds of quantities—for example, miles per hour or dollars per pound. The rate "miles per hour" gives distance traveled per unit of time, and problems using this type of unit rate can be solved using proportions. Students will graph proportional relationships and understand the unit rate informally as a measure of the steepness of a related line—called *slope*. Students distinguish proportional relationships from other relationships.

Additionally, students use their understanding of ratios and proportionality to solve a variety of percent problems. They will solve problems that include simple interest, taxes, tips, commissions, mark-ups, discounts, and percent increases or decreases.

The Grade 7 Common Core State Standards for Ratios and Proportional Relationships specify that students should–

- Analyze proportional relationships and use them to solve real-world and mathematical problems.

The following hands-on activities will help students explore the concepts of ratios and proportional relationships in a meaningful way. The concrete experiences that the activities provide will strengthen students' ability to recognize and work flexibly with these concepts.

Ratios and Proportional Relationships

Contents

Objective

Determine whether a relationship is proportional by checking for a straight-line graph.

Common Core State Standards

- **7.RP.2a** Decide whether two quantities are in a proportional relationship, e.g., by testing for equivalent ratios in a table or graphing on a coordinate plane and observing whether the graph is a straight line through the origin.

Proportional Relationships I

Proportional relationships can be shown arithmetically, graphically, or algebraically. Students at this level will determine if two quantities are in a proportional relationship by graphing their values to see if they form a straight line through the origin. In the future, they will use the equation of a line, $y = mx + b$, to express a proportional relationship.

Try It! *Perform the Try It! activity on the next page.*

Talk About It

Discuss the Try It! activity.

- **Ask:** *What does the x-axis represent?* (time in minutes) *What does the y-axis represent?* (distance in kilometers) *What is the scale on each axis?* (x-axis: one space equals 10 minutes; y-axis: one space equals one kilometer)
- **Say:** *Two points always define a line. Some lines represent proportional relationships and some do not. A line that represents a proportional relationship goes through the origin.*
- **Ask:** *Does the line go through the origin?*

Solve It

Reread the problem with the students. Have students set up a graph and plot the points on the Centimeter Grid (BLM 1). Ask them to use a ruler to draw a line through the points and show that the line goes through the origin.

More Ideas

For other ways to teach about proportional relationships and straight-line graphs—

- Have students work a similar problem that starts with just one point: that a mackerel swims 22 km in 2 hours. Ask them to make a line showing other points that are proportional using the XY Coordinate Pegboard or the Centimeter Grid (BLM 1). Then have them list the points that form a proportional relationship.
- Have students start by making a straight line on their XY Coordinate Pegboard that will pass through the origin. Have them describe all of the points that make the proportional relationship of the line.

Formative Assessment

Have students try the following problem.

Which graph represents a proportional relationship?

A. **B.** **C.** **D.**

Try It! 20 minutes | Pairs

Here is a problem about proportional relationships.

A herring swims 3 kilometers in 30 minutes. Another day, the herring swims 7 kilometers in 70 minutes. Is this a proportional relationship?

Introduce the problem. Then have students do the activity to solve the problem. Distribute the materials.

Materials
- XY Coordinate Pegboard
- BLM 1
- ruler

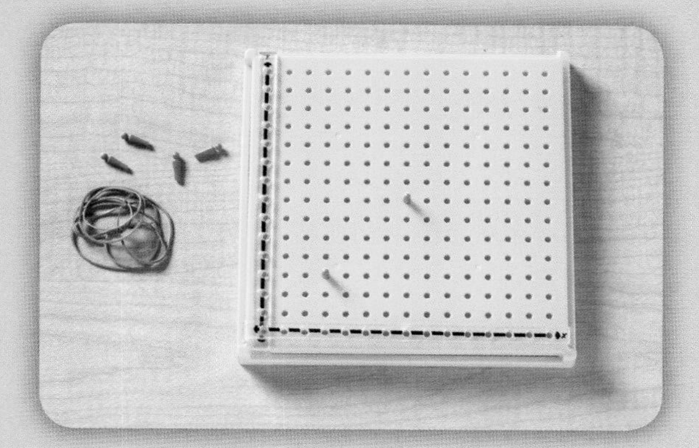

1. Have students set the axes to show the first quadrant of a coordinate plane. **Say:** *The x-axis will represent time in minutes, and we'll have 1 space equal 10 minutes. The y-axis will represent distance, and we'll have 1 space equal 1 kilometer.* Then have students plot the points (30, 3) and (70, 7) on their XY Coordinate Pegboards.

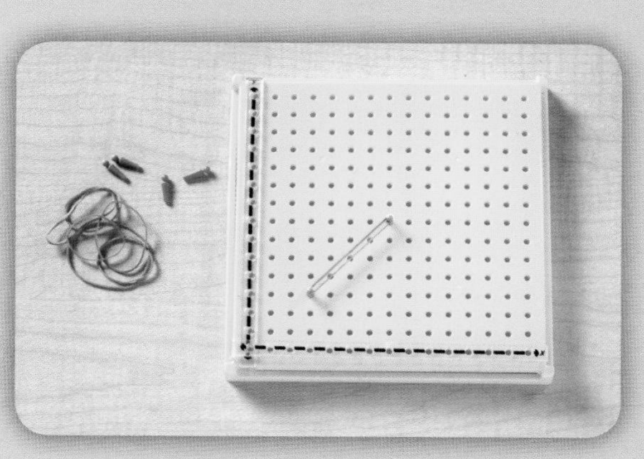

2. Ask students to connect the points with a rubber band to make a line segment on the pegboard. Remind students that any two points define a line.

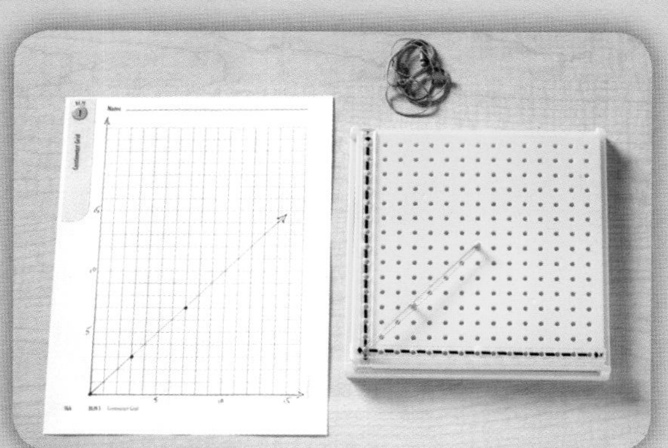

3. Ask: *Where is the origin?* Elicit that the origin is (0, 0). Encourage students to extend their lines to the origin. **Ask:** *Are all three points on a straight line?* Explain that since the points (30, 3) and (70, 7) lie on a straight line that goes through the origin, they express a proportional relationship.

Look Out!

Students might think that any straight line indicates a proportional relationship. Emphasize that the line must go through the origin for the data to express a proportional relationship.

Use an XY Coordinate Pegboard. Complete the model to answer the question.

(Check students' work.)

1. You can buy 3 pounds of bananas for $2 or 9 pounds for $6. Is the relationship proportional?

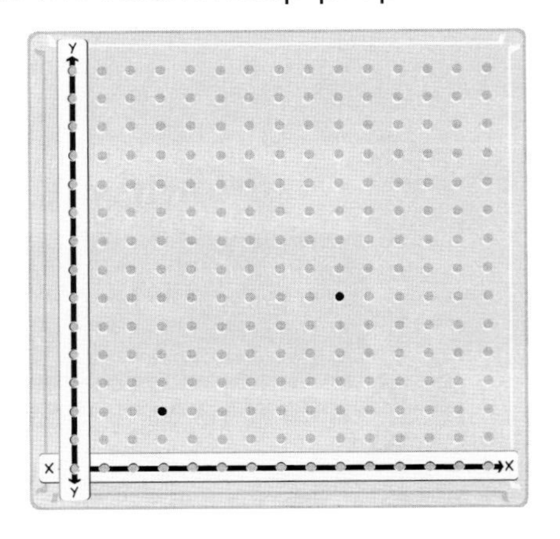

Yes

Using an XY Coordinate Pegboard, model the problem. Draw the model and use it to answer the question.

2. In a bag, there are 4 red balls and 6 blue balls. In a second bag, there are 12 red balls and 8 blue balls. Is the relationship proportional?

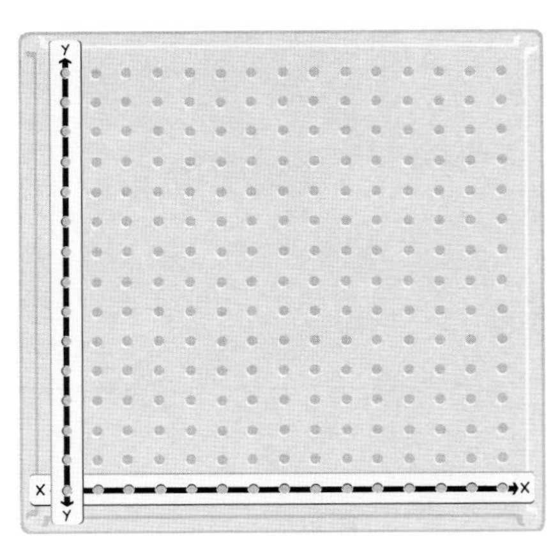

No

Use Centimeter Grid Paper to determine if the relationship is proportional.

3. A baseball player got 14 hits in 35 turns at bat and 32 hits in 80 turns. Is the relationship proportional?

Yes

4. If 20 people are ahead of you in the lunch line, it takes 12 minutes to get your lunch. If 30 people are ahead of you, it takes 18 minutes. Is the relationship proportional?

Yes

5. It rained 15 times in 40 days, and it rained 45 times in 100 days. Is the relationship proportional?

No

Answer Key

Challenge! Explain how you make a graph to determine if some data are in a proportional relationship. Make up an example of a proportional relationship.

Challenge: Graph each data point on a coordinate grid. Draw a line through the points. If the straight line goes through the origin, then the relationship is proportional.

Objective

Determine whether a relationship is proportional by checking for equivalent ratios.

Common Core State Standards

- **7.RP.2a** Decide whether two quantities are in a proportional relationship, e.g., by testing for equivalent ratios in a table or graphing on a coordinate plane and observing whether the graph is a straight line through the origin.

Proportional Relationships II

A proportion is an equation that sets two ratios equal to each other. If that equation is true, then the relationship is proportional. Students have checked to see if relationships are proportional by graphing, but now they will check by using their skills with equivalent fractions.

Try It! *Perform the Try It! activity on the next page.*

Talk About It

Discuss the Try It! activity.

- **Ask:** *What was the first ratio, or fraction, you built?* (2:8, or $\frac{2}{8}$) *What does the ratio represent?* (the number of green apples to the total number of apples) *What was the second ratio, or fraction, you built?* (1:4, or $\frac{1}{4}$) *What does the ratio represent?* (the number of green apples to the total number of apples)
- **Ask:** *How can you tell if the two ratios are equivalent?* (you can build the fractions and compare or simplify)
- **Ask:** *Is this a proportional relationship? Why or why not?* (yes; the ratios are equivalent)

Solve It

Reread the problem with the students. Have students build the two ratios and draw them on the Fraction Squares BLM. Ask students to explain whether the ratios are equivalent and write an equation to represent the equivalent ratios. (2:8 = 1:4, or $\frac{2}{8} = \frac{1}{4}$)

More Ideas

For other ways to teach about proportional relationships and equivalent ratios—

- Have students use Fraction Tower® Equivalency Cubes to build each ratio. Then, they can compare the heights of the towers to see if they are equivalent and therefore represent a proportional relationship.
- Have students make the fractions using Deluxe Rainbow Fraction® Circles. They can measure the fractions with Rainbow Fraction Circle Rings or compare by stacking to determine whether the fractions are equivalent and therefore represent a proportional relationship.

Formative Assessment

Have students try the following problem.

In Rob's group, there are 2 boys and 3 girls. In Caren's group, there are 4 boys and 6 girls. Which equation shows that the groups are proportional?

A. $\frac{2}{6} = \frac{1}{3}$ **B.** $\frac{2}{5} = \frac{4}{10}$ **C.** $\frac{2}{4} = \frac{3}{6}$ **D.** $\frac{2}{3} = \frac{4}{6}$

Try It! 15 minutes | Groups of 4

Here is a problem about proportional relationships.

In a bag of 8 apples, 2 of the apples are green. In a bag of 4 apples, 1 is green. Is this a proportional relationship?

Introduce the problem. Then have students do the activity to solve the problem. Distribute the materials.

Materials

- Deluxe Rainbow Fraction® Squares
- BLM 2

1. Ask: *In the first bag, what is the ratio of green apples to all apples?* Have students use Fraction Squares to represent the ratio 2:8, or $\frac{2}{8}$, on their Fraction Squares BLM.

2. Ask: *In the second bag, what is the ratio of green apples to all apples?* Have students use Fraction Squares to represent the ratio 1:4, or $\frac{1}{4}$, on their Fraction Squares BLM.

3. Ask: *How can you tell if these ratios are equivalent?* Encourage students to stack the fractions to show they are equivalent. Explain that since the ratios are equivalent, the relationship is proportional.

⚠ Look Out!

Students might orient the Fraction Squares differently and therefore think they are not equivalent. Encourage students to align the Fraction Squares pieces vertically and start in the top left corner each time.

Use Fraction Squares. Complete the model to answer the question. (Check students' work.)

1. In the teacher's pencil jar, there are 10 pencils, 4 of which do not have an eraser. In Julio's pencil bag, there are 5 pencils, 3 of which do not have an eraser. Is the relationship proportional?

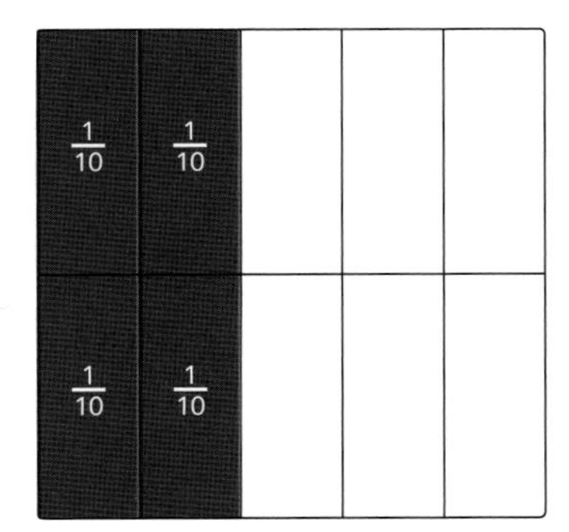

No

Using Fraction Squares, model the problem. Draw the model and use it to answer the question.

2. In a bag of 4 instruments, 2 instruments are shakers. In a box of 8 instruments, 4 are shakers. Is the relationship proportional?

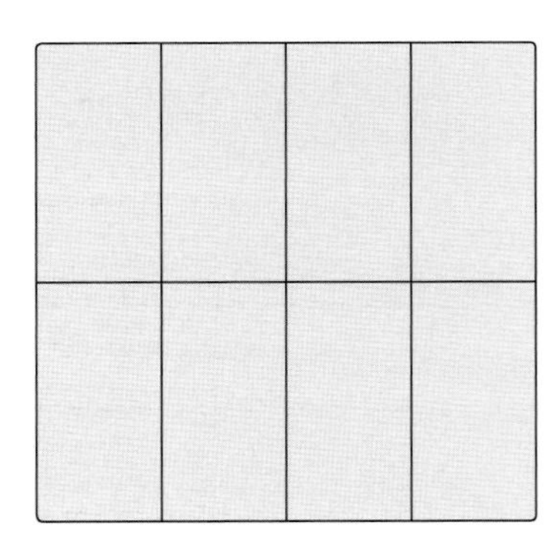

Yes

Use Fraction Squares to determine if the relationship is proportional.

3. In PE, 5 of every 6 girls finished a run in less than 10 minutes. Two of every 3 boys finished in less than 10 minutes. Is the relationship proportional?

No

4. Sal paid $2 for 4 pounds of grapes. Bo paid $1 for 2 pounds of grapes. Is the relationship proportional?

Yes

Use equivalent ratios to determine if the relationship is proportional.

5. Roberto can ride his bike 4 miles in 20 minutes. Patricia can ride her bike 10 miles in 50 minutes. Is the relationship proportional?

Yes

6. Pearla answered 4 of the 5 questions right on the quiz. Then, on the test, she answered 15 of the 20 questions right. Is the relationship proportional?

No

Answer Key

Challenge! Ms. Turny's class ratio of boys to girls is 2:3, and Mr. Straight's class ratio of boys to girls is 8:12. Explain how you know if the data portrays a proportional relationship.

Challenge: If the data forms equivalent ratios, then the relationship is proportional. Since 2:3 = 8:12, the data does portray a proportional relationship.

Objective

Identify the constant of proportionality for a proportional relationship.

Common Core State Standards

- **7.RP.2b** Identify the constant of proportionality (unit rate) in tables, graphs, equations, diagrams, and verbal descriptions of proportional relationships.
- **7.RP.2d** Explain what a point (x, y) on the graph of a proportional relationship means in terms of the situation, with special attention to the points (0, 0) and $(1, r)$ where r is the unit rate.

Ratios and Proportional Relationships

Constant of Proportionality

A relationship is proportional if its graph is a straight line through the origin. The slope of this line is the constant of proportionality, or unit rate, for the relationship. The unit rate is also revealed on the line by the point $(1, r)$, where r is the unit rate. Graphing a proportional relationship is the most powerful way to help students visualize what the constant of proportionality represents.

Try It! *Perform the Try It! activity on the next page.*

Talk About It

Discuss the Try It! activity.

- **Say:** *The problem says the number of blocks is proportional to the number of minutes, so the number of blocks goes on the y-axis and time goes on the x-axis.*
- Explain that the calculation of the constant of proportionality will produce the same result regardless of the two points chosen from the line. **Ask:** *Which point on the line will give you the constant of proportionality directly, without having to calculate?* Elicit that the answer is the point whose x-coordinate is 1. That point is $(1, r)$, where r is the unit rate, or constant of proportionality.

Solve It

Reread the problem with the students. Have students show the graph on grid paper. Ask them to find the vertical distance and the horizontal distance between the two given points and divide to find the constant of proportionality. Help them recognize that the constant is the y-value that the line crosses at $x = 1$.

More Ideas

For another way to teach about the constant of proportionality—

- Have students use Fraction Tower® Equivalency Cubes to solve the problem. Start by asking students to build a whole tenths tower and a whole fifths tower. Have students lay down the fifths tower horizontally and then lay down the tenths tower horizontally immediately below it. Tell students the towers represent the ratio 5:10 (5 blocks in 10 minutes) because there are 5 fifths and 10 tenths. Next have students break the towers to show the ratio 2:4, then 1:2. Elicit from students that to find the unit rate, they need to find the number of green cubes for each purple cube. Help students see that they need $\frac{1}{2}$ of a green cube for each purple cube and that this $\frac{1}{2}$ is the unit rate, or constant of proportionality in the problem.

Formative Assessment

Have students try the following problem.

Teo has 3 trophies and 6 medals. Brandi has 5 trophies and 10 medals. If the number of medals is proportional to the number of trophies, which point will be on the graph of this relationship?

A. (0, 1) **B.** (1, 2) **C.** (4, 7) **D.** (12, 5)

Try It! 15 minutes | Pairs

Here is a problem about proportional relationships and the constant of proportionality.

Liam took his dog for a walk. In 4 minutes, he had walked 2 blocks. In 10 minutes, he had walked 5 blocks. If the number of blocks is proportional to the number of minutes, what is the constant of proportionality for the relationship?

Introduce the problem. Then have students do the activity to solve the problem. Distribute the materials.

Materials

- XY Coordinate Pegboard
- BLM 1

1. Say: *Let's graph the relationship on the pegboard. Put distance on the y-axis and time on the x-axis.* **Ask:** *What two points will you graph?* Have students place pegs at (4, 2) and (10, 5) and connect the points with a rubber band. **Ask:** *Can you confirm that the relationship is proportional?* Ask students to extend their bands to the origin. Elicit that the relationship is proportional.

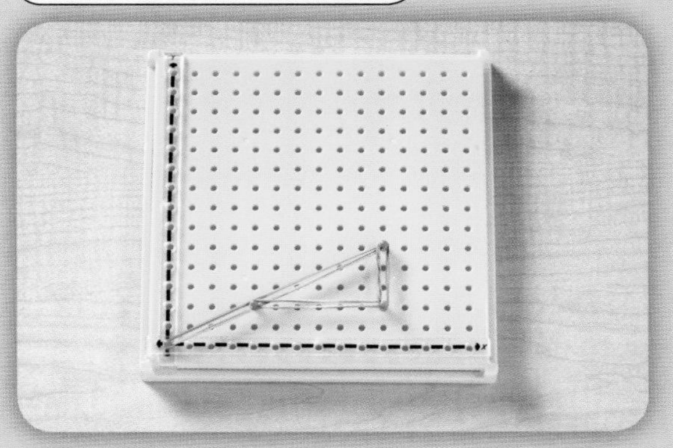

2. Ask: *What is the horizontal distance between the two points?* Have students use a rubber band and peg to mark the horizontal distance between (4, 2) and (10, 5). **Ask:** *What is the vertical distance between the two points?* Have students use another band to mark the vertical distance between (4, 2) and (10, 5).

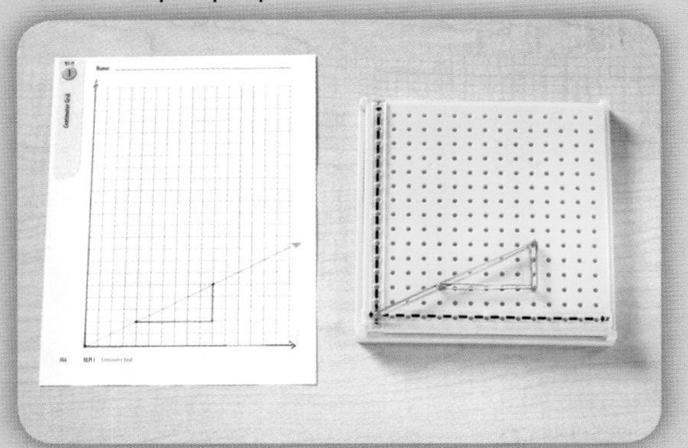

3. Say: *Divide the vertical distance by the horizontal distance. The quotient is called the* constant of proportionality. *It is also called* the unit rate, *because it tells how much the y-coordinate changes for each unit of change in the x-coordinate*. Have students confirm the unit rate by inspecting the graph.

⚠ Look Out!

Students might have difficulty finding the vertical and horizontal distances between points. Point out that the rubber bands that represent the distances form a right triangle with the line.

Use an XY Coordinate Pegboard. Build the model and use it to answer the question.

(Check students' work.)

1. For every 2 apples in Kali's orchard, there are 4 pears. In Sam's orchard, there are 10 pears for every 5 apples. If the number of pears is proportional to the number of apples, what is the constant of proportionality?

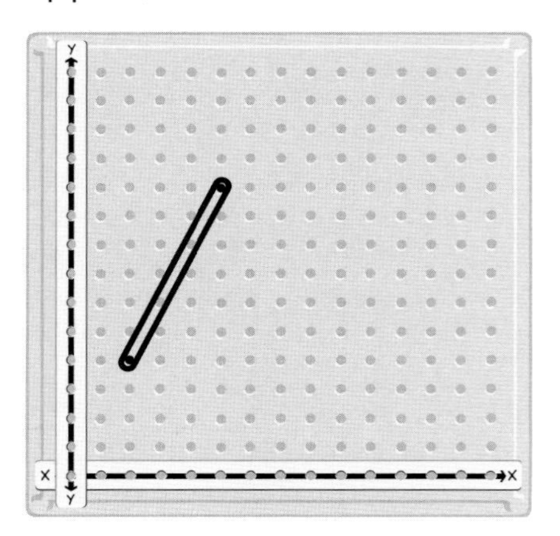

2

Using an XY Coordinate Pegboard, model the problem. Draw the model and use it to answer the question.

2. Yesterday, Maria had 4 nickels and 3 dimes in her wallet. Today, she has 12 nickels and 9 dimes. If the number of dimes is proportional to the number of nickels, what is the constant of proportionality?

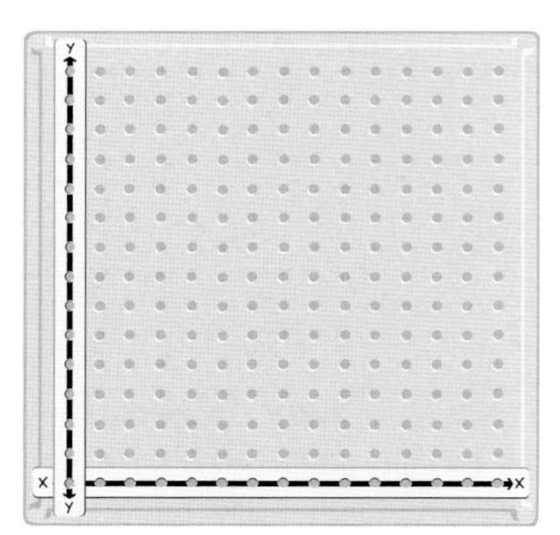

$\frac{3}{4}$

Use Centimeter Grid Paper to find the constant of proportionality for the relationship.

3. Carson can jump 30 times in 18 seconds. He can jump 20 times in 12 seconds. Assume that the number of times Carson can jump is proportional to the number of seconds he is given.

$\frac{5}{3}$

4. Cleo bought 36 stamps and paid $12. Ned paid $6 for 18 stamps. Assume the amount paid for stamps is proportional to the number of stamps purchased.

$\frac{1}{3}$

5. Sheila made 14 three-point shots in 35 attempts. She made 32 three-point shots in 80 attempts. Assume the number of three-point shots Sheila makes is proportional to her number of attempts.

$\frac{2}{5}$

Answer Key

Challenge! Caitlyn read 2 books in 8 days. She read 4 books in 16 days. Assume that the number of books Caitlyn reads is proportional to the number of days she spends reading. Find the constant of proportionality for the relationship. Identify the point on a graph of the relationship that directly names the value of the constant. Explain why that point works.

Challenge: The two points represented in the problem are (8, 2) and (16, 4). The vertical distance between them is $4 - 2 = 2$. The horizontal distance is $16 - 8 = 8$. The constant of proportionality is $\frac{1}{4}$. The point that directly names the constant is $(1, \frac{1}{4})$. This is so because the constant of proportionality is the change in y for each unit change in x. The point $(1, \frac{1}{4})$ represents a 1-unit change in x relative to the point (0, 0). The vertical change is $\frac{1}{4} - 0 = \frac{1}{4}$ and the horizontal change is $1 - 0 = 1$, so the y-coordinate of the point $(1, \frac{1}{4})$ literally names the constant of proportionality.

Equations of Proportional Relationships

Objective

Develop a set of coordinate pairs for a proportional relationship; write and graph an equation for the relationship.

Common Core State Standards

- **7.RP.2b** Identify the constant of proportionality (unit rate) in tables, graphs, equations, diagrams, and verbal descriptions of proportional relationships.
- **7.RP.2c** Represent proportional relationships by equations. *For example, if total cost t is proportional to the number n of items purchased at a constant price p, the relationship between the total cost and the number of items can be expressed as $t = pn$.*

In previous lessons, students have learned about proportional relationships and have learned to find the constant of proportionality for a relationship using coordinate pairs. These coordinate pairs are related linearly, meaning that the graph of their relationship forms a straight line. In this activity, students will use a set of related coordinate pairs to graph and write the equation for a proportional relationship.

Try It! *Perform the Try It! activity on the next page.*

Talk About It

Discuss the Try It! activity.

- **Ask:** *Why aren't we pegging the number of people that Kelly meets each day?*
- **Ask:** *How do you know the graph shows a proportional relationship? Could we have known this without graphing the coordinates? Explain.*
- **Ask:** *What is the constant of proportionality?*

Solve It

Ask students to explain why Kelly's plan to meet three new people every two days is a proportional relationship. Encourage them to use their graph to explain their answer. Have students write the equation for the relationship.

More Ideas

For another way to teach about proportional relationships and equations—

- Give students a related set of ordered pairs. Have students use their XY Coordinate Pegboard to determine whether the set of ordered pairs defines a proportional relationship. Have them use the graph to explain their answer.

Formative Assessment

Have students try the following problem.

Which of the following lists of values shows a proportional relationship?

A.

x	y
1	1
2	4
3	9
4	16
5	25

B.

x	y
4	2
8	4
12	6
16	4
24	2

C.

x	y
0	−5
1	−4
2	−3
3	−4
4	−1

D.

x	y
0	0
3	6
6	12
9	18
12	24

Try It! 20 minutes | Pairs

Here is a problem about proportional relationships and equations.

Kelly loves to meet new people. When she moved to a new school, she decided to meet three new people every two days. How many people will she have met after 10 days? After 16 days? Write an equation for the number of people Kelly will have met after x days.

Introduce the problem. Then have students do the activity to solve the problem. Distribute the materials.

Materials

- XY Coordinate Pegboard
- paper

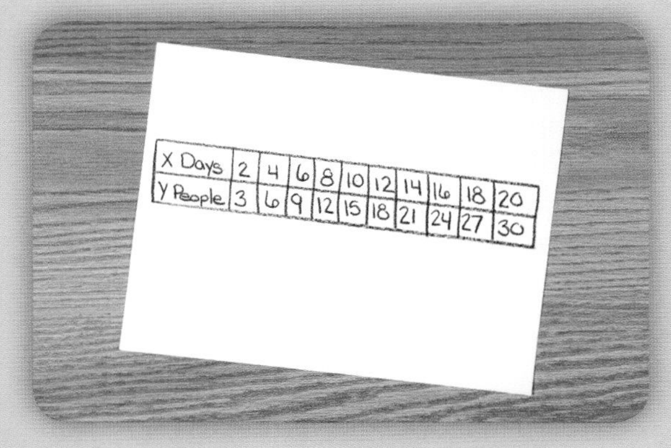

X Days	2	4	6	8	10	12	14	16	18	20
Y People	3	6	9	12	15	18	21	24	27	30

1. Have students set up a table of values showing each day and the number of new people Kelly has met by the end of that day. Have them start at $x = 2$ and continue through $x = 20$.

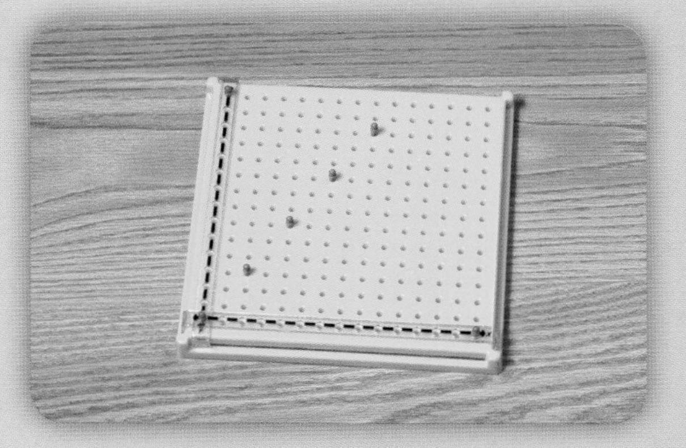

2. Have students peg the first few sets of coordinate pairs from their table of values.

3. Have students find the constant of proportionality. Tell them to look for a pattern. **Ask:** *What is the rule for moving from one point on the graph to the next?* Elicit from them that the pattern is "up 3, over 2."

4. Have students write their solutions on a sheet of paper. Their formulas should be in the form $y = mx$.

Use an XY Coordinate Pegboard to plot the points shown. Make a table of ordered pairs. Graph the line. Write an equation. (Check students' work.)

1.

x	y
0	0
6	4
9	6
12	8
15	10
18	12

$y = \frac{2}{3}x$

Using an XY Coordinate Pegboard, graph the line that passes through the points given on the grid. Sketch the line. Make a table of ordered pairs. Write an equation.

2.

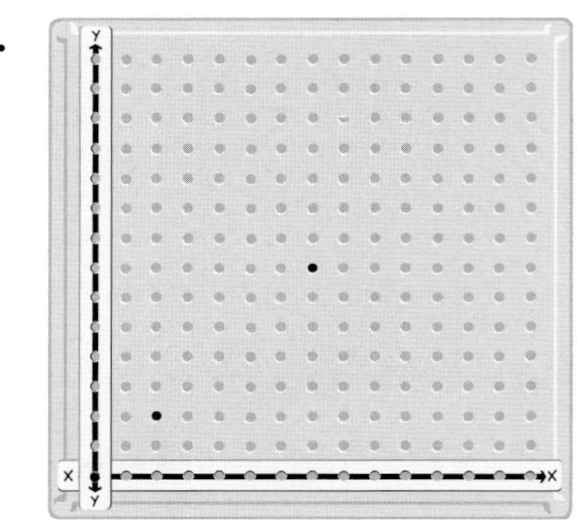

x	y
0	0
2	2
4	4
7	7
9	9
11	11

$y = x$

Graph a line that passes through the given points. Make a table of ordered pairs. Write an equation.

3.

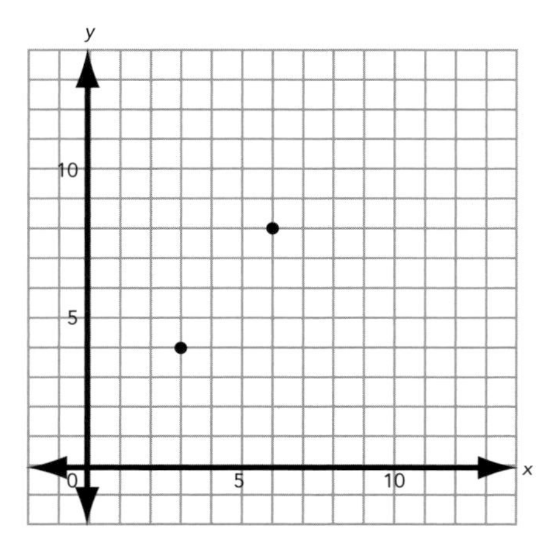

x	y
0	0
3	4
6	8
9	12
12	16
15	20

$y = \frac{4}{3}x$

Answer Key

Challenge! How many points must you have to make a line? Why is it good to have three points to make a line?

Challenge: (Sample) Two points make a line. When you graph three points on a line, it is a good way to check that the first two points are graphed accurately.

The Number System

In seventh grade, students develop greater understanding of the number system by exploring **rational numbers**: integers, fractions, decimals, and percents.

At this level, students apply and extend previously learned concepts of addition, subtraction, multiplication, and division to adding, subtracting, multiplying, and dividing with any of the rational numbers. For instance, students may find the sum -10 + 5 by locating -10 on a number line, moving 5 spaces in the positive direction, and interpreting the number they land on (-5) as the sum. Similarly, students apply long-division concepts to learn the difference between terminating and repeating decimals—a concept necessary for later work with rational and irrational numbers.

Students continue building on their previous work with the order of operations to solve problems with rational numbers. The **order of operations** is a set of rules for determining the order in which the operations in an expression are performed. Students will apply their expanded view of the rational numbers when they work with algebraic expressions. They will learn that rewriting an expression in different forms can help them solve problems. They will also learn to work more confidently with negative numbers by viewing and experiencing them in everyday contexts.

The Grade 7 Common Core State Standards for The Number System specify that students should—

- Apply and extend previous understandings of operations with fractions to add, subtract, multiply, and divide rational numbers.

The following hands-on activities enable teachers to provide rich opportunities for students to deepen their understanding of the number system, with particular regard to the rational numbers. The experiences will help students to develop a unified understanding of number—that is, to work flexibly with integers, fractions, decimals, and percents.

The Number System

Contents

Objective

Add integers.

Common Core State Standards

- **7.NS.1b** Understand $p + q$ as the number located a distance $|q|$ from p, in the positive or negative direction depending on whether q is positive or negative. Show that a number and its opposite have a sum of 0 (are additive inverses). Interpret sums of rational numbers by describing real-world contexts.

Add Integers I

Addition is the same for whole numbers and integers—the grouping together of quantities. Modeling with different colors helps students to perform operations with positive and negative numbers. As students learn the rules for working with integers, they should make the connection between the models they build and the rules so that manipulating integers is not arbitrary.

Try It! *Perform the Try It! activity on the next page.*

Talk About It

Discuss the Try It! activity.

- **Ask:** *After you model the problem with Two-Color Counters, why do you rearrange them to form red-yellow pairs? What does a red-yellow pair represent in terms of yardage?* Students should recognize that one yard lost (red) plus one yard gained (yellow) is a net change of zero, so a red-yellow pair represents no gain or loss.
- **Ask:** *Is the sum of two negative numbers always negative? Model an example to justify your answer.*

Solve It

Reread the problem with students. Since 12 > 5, the team gained more yards than it lost. Since 12 – 5 = 7, they gained 7 more yards than they lost. So their net yardage is 7 yards. Have students write –5 + 12 = 7 and explain how this equation relates to the problem.

More Ideas

For other ways to teach about adding integers—

- Have students use Centimeter Cubes and a 1-cm Number Line (BLM 3) to add pairs of integers—two positive numbers: 2 + 6, two negative numbers: –1 + (–5), and a positive number and negative number: –9 + 2 or 8 + (–4). Suggest that students use red cubes for negative numbers and yellow cubes for positive numbers.
- Ask students to write their own number sentences with positive and negative numbers and use Color Tiles to model the sentences and show that a positive number plus a negative number can be positive, negative, or zero.

Formative Assessment

Have students try the following problem.

The morning temperature of –9°F is expected to rise 10 degrees by noon. What is the expected noon temperature?

A. –19°F **B.** –1°F **C.** 1°F **D.** 19°F

Try It! 15 minutes | Pairs

Here is a problem about adding integers.

A football team lost 5 yards on one play and then gained 12 yards on the next play. What was the team's net yardage on the two plays?

Introduce the problem. Then have students do the activity to solve the problem. Distribute the materials.

Materials
- Two-Color Counters (at least 20 per pair)
- BLM 3

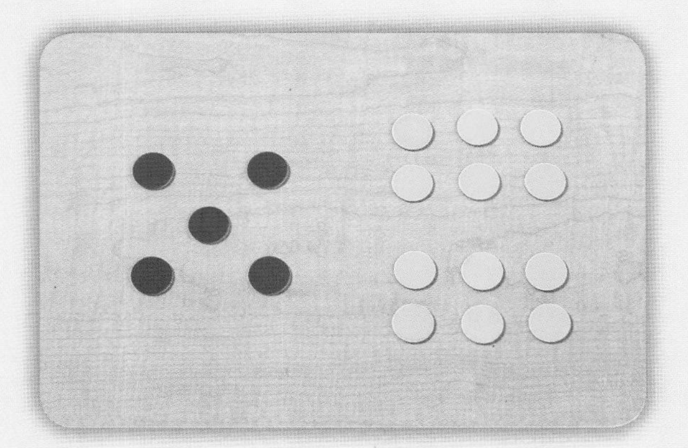

1. Say: *Each red counter represents one yard lost. Each yellow counter represents one yard gained. Place counters to model this problem.* Students place 5 red counters and 12 yellow counters.

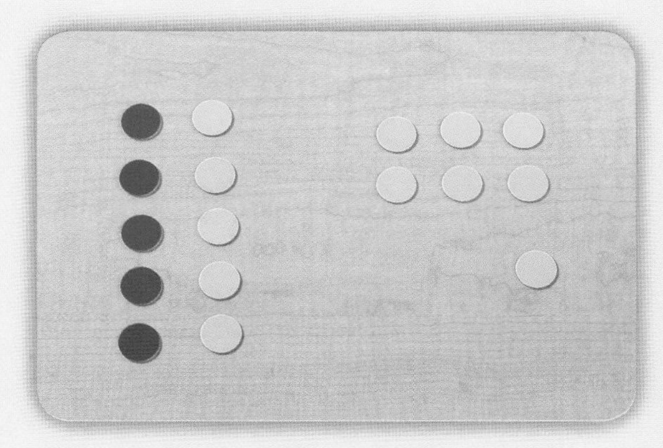

2. Say: *Move counters so that each red counter is paired with a yellow counter.*
Ask: *What number does each red-yellow pair represent? How many yellow counters are left?* Students form 5 pairs, each representing 0. There are 7 yellow counters left, representing a net gain of 7 yards.

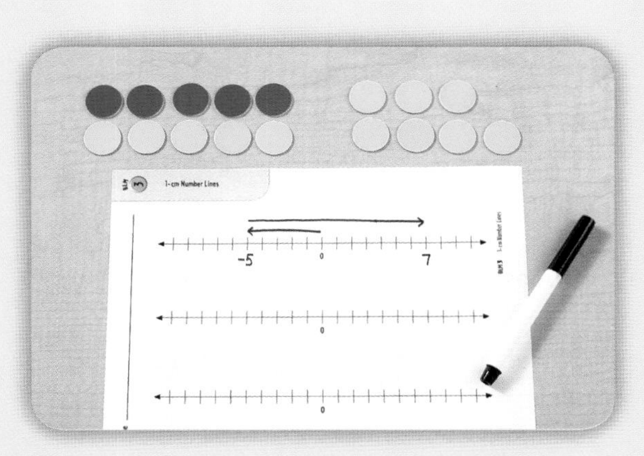

3. Say: *Now use a number line to solve this problem. Starting at 0, draw a segment 5 units to the left. From –5, draw a segment 12 units to the right.* **Ask:** *At what number do you end?* Students draw the two segments and end at 7. Help students recognize that the overlapping parts of the lines are equivalent to red-yellow pairs of counters.

⚠ Look Out!

Students may confuse negative signs with minus signs. Have students write the problem –5 + 12 = 7. Then speak the correct words: *negative five* (not *minus five*) *plus twelve equals seven.* On a number line, show students that –5 and 5 are opposites. They are both 5 units from 0, but in opposite directions.

Use Two-Color Counters to model each addition problem. Make pairs of red and yellow counters. Find the sum. (Check students' work.)

1.  

9 + (–10)

–1

2. 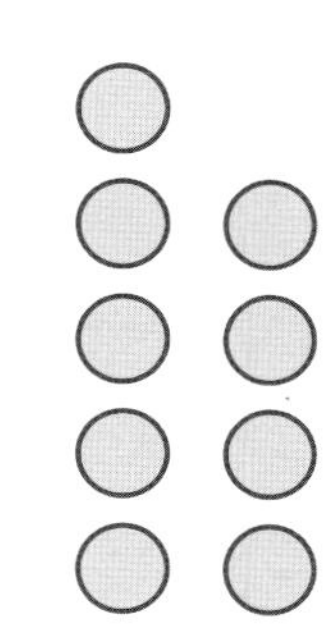

–18 + 9

–9

Using Two-Color Counters, model each addition problem. Sketch the model. Find the sum.

3. 7 + (–4)

3

4. –12 + (–3)

–15

Find each sum.

5. 11 + (–6) 5

6. –5 + (–18) –23

7. –4 + 13 9

8. 9 + (–21) –12

9. –6 + (–14) –20

10. –3 + (–18) –21

11. 15 + 9 24

12. 26 + (–50) –24

Answer Key

Challenge! Explain how to add two integers if one is a negative number and one is a positive number. When will the sum be negative? When will the sum be positive?

Challenge: (Sample) You actually subtract the numbers, ignoring the signs, and the answer will have the same sign as the number with the greater value (ignoring the negative signs).

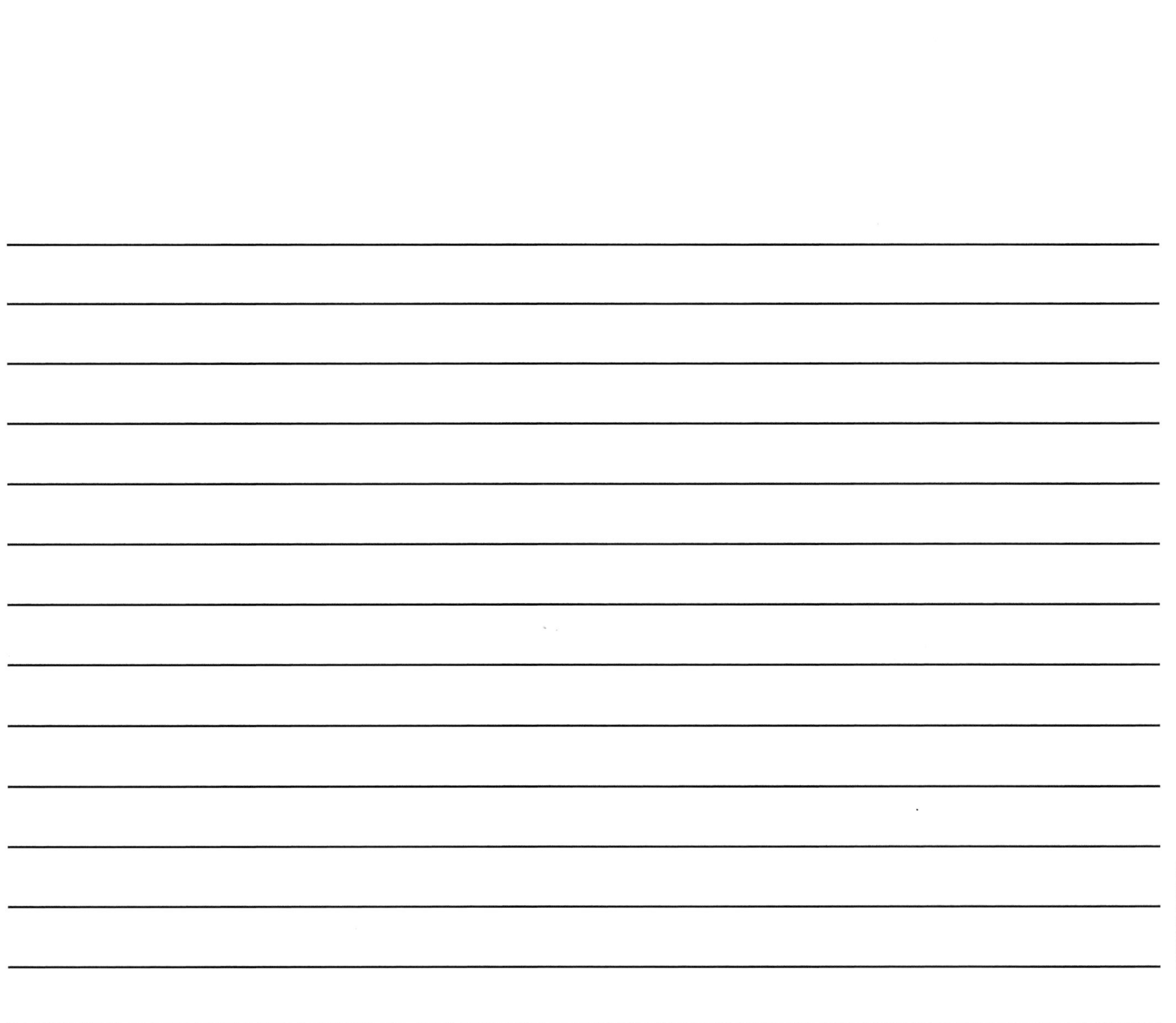

Objective

Add integers.

Common Core State Standards

- **7.NS.1b** Understand $p + q$ as the number located a distance $|q|$ from p, in the positive or negative direction depending on whether q is positive or negative. Show that a number and its opposite have a sum of 0 (are additive inverses). Interpret sums of rational numbers by describing real-world contexts.

Add Integers II

Students may be familiar with integer concepts in everyday situations such as temperatures above and below zero, altitudes above and below sea level, and football yardage gained and lost. Students should recognize zero pairs and be able to use the identity property of addition to simplify computation. At the 7th and 8th grade levels, they need to be able to compute with integers in preparation for using integers in solving equations and inequalities.

Try It! *Perform the Try It! activity on the next page.*

Talk About It

Discuss the Try It! activity.

- **Ask:** *What is the opposite of turning on a light? What is the opposite of sitting down? What is the opposite of +1? If we add +1 and its opposite (–1) what do we get?*
- **Ask:** *What does a discount do to the price of an item? How should we represent a discount coupon?*
- **Ask:** *How many zero pairs are there? After you have taken the zero pairs off the mat, what is left on the mat?*

Solve It

Reread the problem with the students. Have them write an explanation of the term "zero pairs." Students should include a discussion of why zero pairs are important in solving the problem.

More Ideas

For other ways to teach the addition of integers—

- Have pairs of students using Two-Color Counters designate one color as positive and the other color as negative. The first student makes up an addition problem with at least one negative addend and the other student models and solves it. Have students trade roles and repeat.
- Suggest that some students redo the Try It! activity with Two-Color Counters.

Formative Assessment

Have students try the following problem.

At 8:00 A.M., the temperature at Arctic base camp was –38°C. By 11:00 A.M. the temperature had risen 17 degrees. What was the 11:00 A.M. temperature?

A. –55°C　　**B.** –45°C　　**C.** –21°C　　**D.** 55°C

Try It! 15 minutes | Pairs

Here is a problem about adding integers.

At the school store, students can use discount coupons earned in class for excellent work and good behavior. The transactions shown in the chart took place today.
How much money did the school store receive today?

Student	Purchase	Price	Coupon
Sally	Pack of Pencils	$1.00	No
Damon	Spiral Notebook	$2.00	$1.00
Rongita	Pack of Pencils	$1.00	$1.00
Chung	Protractor Set	$1.00	No
Dean	Compass Set	$2.00	$1.00
Paul	Glue Stick	$1.00	No

Introduce the problem. Then have students do the activity to solve the problem. Write the chart of transactions on the board. Distribute the materials.

Materials
- Algeblocks® units
- BLM 4

1. Have students read the first transaction from the chart. **Say:** *Put one unit block in the positive section of your Algeblocks Basic Mat. Write +1 on your paper.*

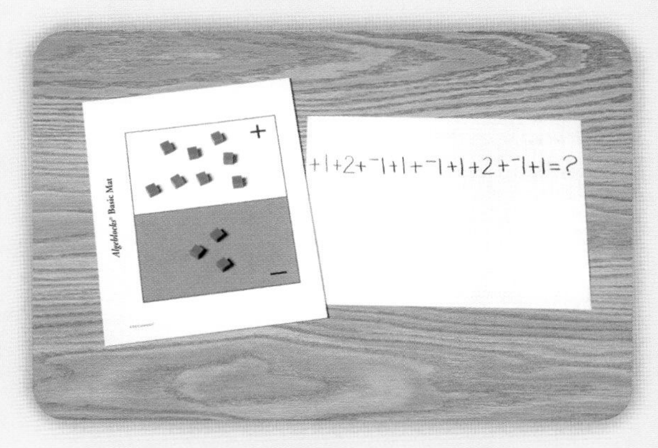

2. Have students read the second transaction. **Say:** *Put two unit blocks in the positive section of your mat, and put one unit block in the negative section. Write +2 and –1 on your paper. Now represent the rest of the transactions.*

3. Say: *Make zero pairs by pairing a unit block in the positive section with a unit block in the negative section. Repeat until all pairs are made. Now count the remaining unit blocks.*

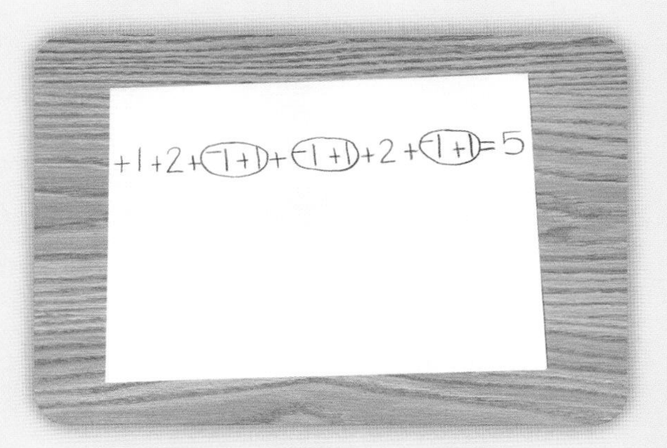

4. Have students circle all of the zero pairs. **Say:** *Add up the remaining numbers. What is the sum? This is the total amount the school store received.*

Use Algeblocks unit blocks and a Basic Mat to model each integer addition sentence. Make zero pairs. Write the sum.

(Check students' work.)

1. $5 + (-8) =$ −3

2. $-3 + 9 =$ 6

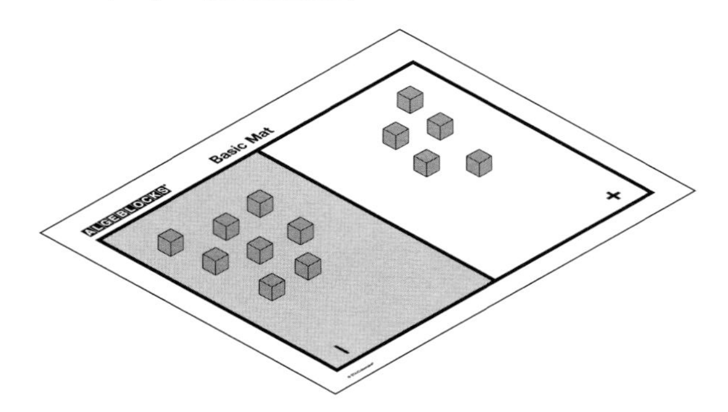

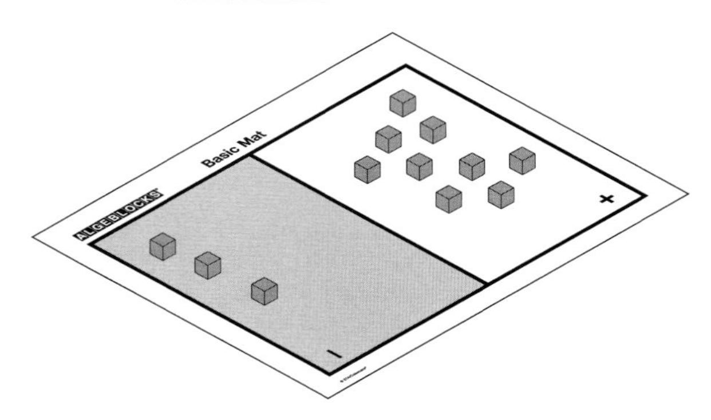

Using Algeblocks unit blocks and a Basic Mat, model each addition sentence. Sketch the model. Circle zero pairs. Write the sum.

3. $12 + (-7) =$ 5

4. $-2 + (-5) =$ −7

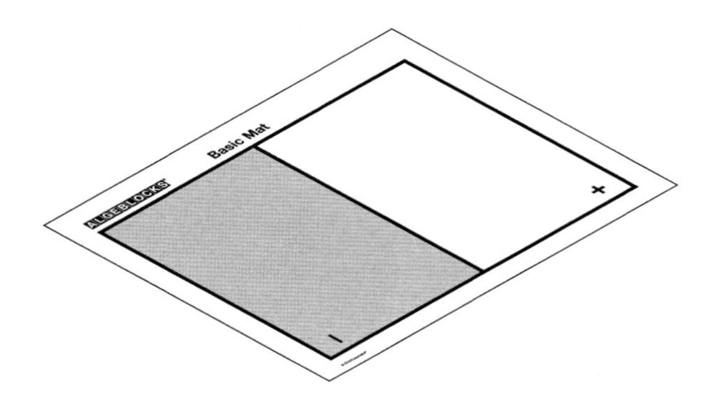

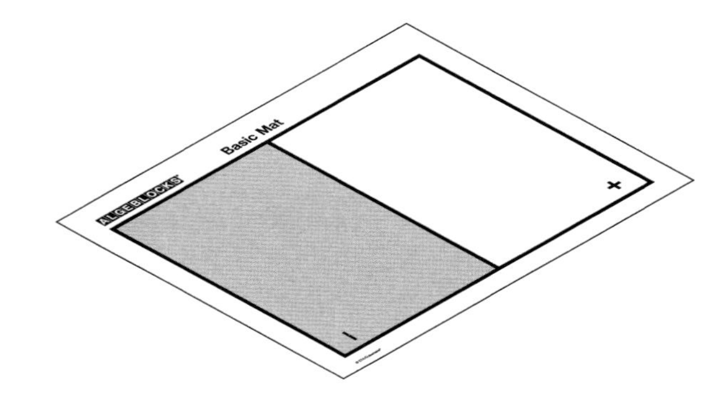

Find each sum.

5. $-5 + (-2) =$ −7

6. $15 + (-4) =$ 11

7. $8 + (-11) =$ −3

8. $-9 + 13 =$ 4

9. $-17 + (-4) =$ −21

10. $15 + 3 =$ 18

11. $-12 + 12 =$ 0

12. $21 + (-7) =$ 14

Challenge! Describe how to find a sum of two integers when the signs of the integers are different. How do you decide the sign of the sum?

Challenge: (Sample) When the signs of the integers are the same, add the digits and use the sign of the numbers as the sign of the sum. When the signs are different, subtract the digits and use the sign of the digit with the greater value as the sign of the sum.

Objective

Subtract integers.

Common Core State Standards

- **7.NS.1c** Understand subtraction of rational numbers as adding the additive inverse, $p - q = p + (-q)$. Show that the distance between two rational numbers on the number line is the absolute value of their difference, and apply this principle in real-world contexts.

Subtract Integers I

After students show confidence with adding integers, they can learn to subtract integers. They will continue to use and develop their understanding of addition and subtraction as inverse operations. Previous work with fact families will help students to think flexibly as they add and subtract positive and negative numbers.

Try It! *Perform the Try It! activity on the next page.*

Talk About It

Discuss the Try It! activity.

- **Ask:** *Is this problem about a subtraction problem or an addition problem?* Discuss with students.
- **Ask:** *When you think about this problem in two different ways—as addition and as subtraction—do you get two different answers?*
- Have students write the two number sentences for this problem.

Solve It

Reread the problem with students. Notice that only red counters are used to solve the subtraction problem, $-8 - (-6) = -2$. Both yellow and red counters are used to solve the addition problem, $-8 + 6 = -2$. Make sure students understand both ways to think about this problem. Either way, Hannah still owes Rachel \$2 at the end.

More Ideas

For other ways to teach about subtracting integers—

- Students can use red and yellow Color Tiles to model the problem.
- Have students use Centimeter Cubes to find $-2 - (-5)$. Suggest that students use red cubes for negative numbers and yellow cubes for positive numbers. They start with 2 red cubes and need to take away 5. But there are only 2 cubes available to take away, so 3 red-yellow pairs (which equal 0) can be added. Then 5 red cubes are removed, and 3 yellow cubes are left.

Formative Assessment

Have students try the following problem.

The current temperature is –6°F and is expected to drop 10 degrees overnight. What is the expected low temperature overnight?

A. –16°F **B.** –10°F **C.** –4°F **D.** 4°F

Try It! 15 Minutes | Pairs

Here is a problem about subtracting integers.

At the bookstore, Hannah borrowed $8 from her sister Rachel. At the waterpark a few days later, Rachel borrowed $6 from Hannah. What is Hannah's standing with Rachel now? Does Hannah still owe Rachel any money?

Introduce the problem. Then have students do the activity to solve the problem. Distribute the materials.

Materials
- Two-Color Counters (at least 20 per pair)

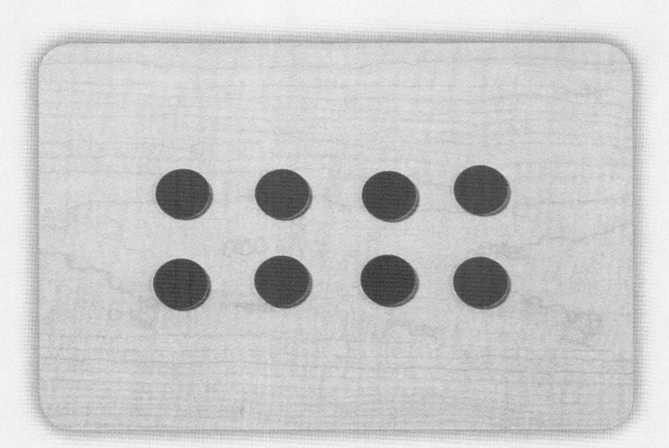

1. Say: *Let each red counter represent one dollar owed, or –1. Use counters to show Hannah's situation after borrowing $8 from Rachel.* Students place 8 red counters on the table.

2. Say: *Later, Rachel borrowed $6 from Hannah. One way to think of this is that $6 of Hannah's debt to Rachel is taken away. This is a subtraction problem: –8 – (–6). Show this with the counters.* Students take away 6 red counters, and 2 are left.

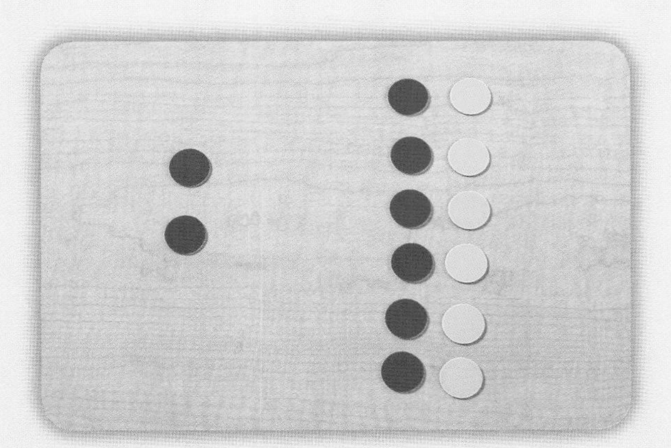

3. Say: *You can also think that when Rachel borrowed $6 from Hannah, it was the same as Hannah paying $6 back to Rachel. It is an addition problem: –8 + 6. Show this with the counters.* Students place 8 red counters, then add 6 yellow counters. They form 6 red-yellow pairs, and 2 red counters are left.

⚠ Look Out!

Students often get confused when they try to subtract a negative number, as in –8 – (–6). When they *take away* 6 red counters from a set of 8, students see that they can actually subtract a negative number. In this activity they also see that subtracting negative 6 is the same as adding positive 6: –8 – (–6) = –8 + 6. Once students are convinced of this, encourage them to use this concept whenever they see a minus sign and a negative sign together. For example, 1 – (–4) = 1 + 4 = 5.

Use Two-Color Counters to model each subtraction problem. Write the number sentence for the difference. (Check students' work.)

1.

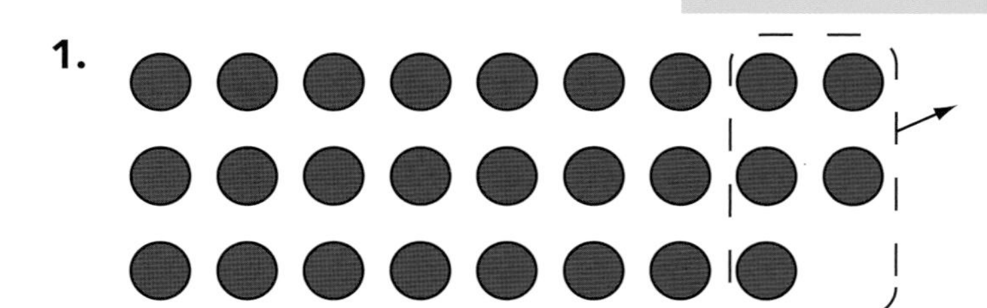

−26 − (−5) = −21

2.

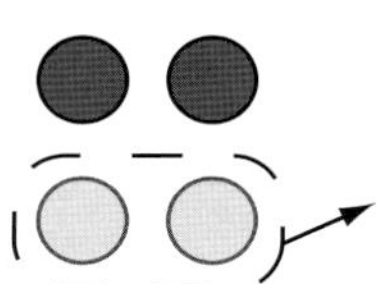

−15 − 2 = −17

Using Two-Color Counters, model each subtraction problem. Sketch the model. Find the difference.

3. 9 − (−4)

13

4. −14 − 5

−19

Find each difference.

5. 21 − (−6) 27

6. −15 − 7 −22

7. −4 − 12 −16

8. −9 − (−7) −2

Answer Key

Challenge! Rewrite Questions 5–8 as addition problems. Find the sum. Did your answers change? Explain.

Challenge: (Sample) 21 + 6 = 27; –15 + (–7) = –22; –4 + (–12) = –16; –9 + 7 = –2; No; Adding the opposite of a number is the same as subtracting a number.

Objective

Subtract integers.

Common Core State Standards

- **7.NS.1c** Understand subtraction of rational numbers as adding the additive inverse, $p - q = p + (-q)$. Show that the distance between two rational numbers on the number line is the absolute value of their difference, and apply this principle in real-world contexts.

The Number System

Subtract Integers II

The addition of integers is straightforward and fairly easy for students to perform once they understand how to use the identity property of addition to make zero pairs. Now students add zero pairs to allow them to subtract quantities that result in a remainder of less than zero. Some students may recognize that this technique is similar to the renaming that they did when they learned to subtract a larger digit from a smaller digit in second or third grade.

Try It! *Perform the Try It! activity on the next page.*

Talk About It

Discuss the Try It! activity.

- **Ask:** *What does "Dropped 5°" mean in this context? Can you take away 5 unit blocks?*
- **Ask:** *Can you take away 3 unit blocks? How many zero pairs would we need to add until we have the 3 unit blocks we need to take away?*
- **Ask:** *Did you all end up with the correct noon reading of –5°? How many zero pairs did you have to add for the final temperature change?*

Solve It

Reread the problem with students. Ask the students to make a chart similar to the one in the story problem and repeat the entire process, recording the temperature changes as they work through the activity.

More Ideas

For another way to teach about subtraction of integers—

- Have pairs of students use polyhedral dice as number generators to create integer subtraction problems. The first student designates both dice as positive or as negative or one die as positive and the other as negative. The first student rolls the dice and writes a subtraction problem using the numbers thrown. The second student models the problem on his or her Algeblocks® Basic Mat and gives the answer. The first student checks the answer. Have students trade roles and repeat.

Formative Assessment

Have students try the following problem involving subtraction of integers.

The early morning temperature reading was 9°C. By noon, the temperature had dropped 15°. What was the temperature at noon?

A. –24°C **B.** –6°C **C.** 6°C **D.** 24°C

Try It! 15 minutes | Pairs

Here is a problem about the subtraction of integers.

Ann's class is recording changes in temperature for science class. Students check the temperature readings each hour and determine the change in temperature. The chart shows the changes. What were the readings?

Time	8 A.M.	9 A.M.	10 A.M.	11 A.M.	Noon
Reading	+6°				
Change		Dropped 5°	Dropped 3°	Dropped 1°	Dropped 2°

Introduce the problem. Then have students do the activity to solve the problem. Write the chart on the board. Distribute the materials.

Materials
- Algeblocks® units
- BLM 4

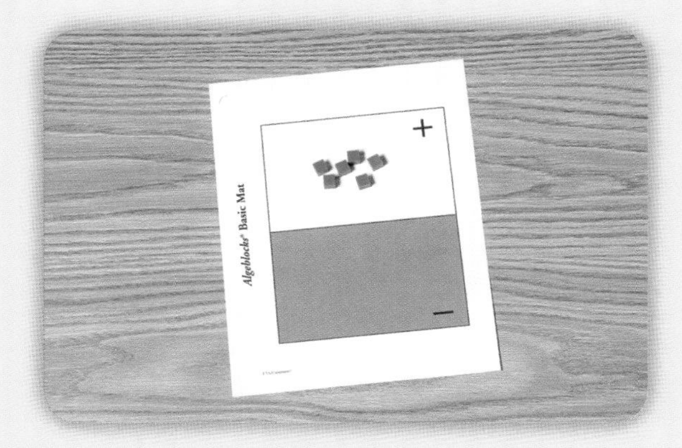

1. Have students place 6 unit blocks on the positive side of the mat. **Say:** *Read the first change in temperature, "Dropped 5°." Take away 5 cubes.* **Ask:** *What was the 9 A.M. reading?*

2. Have students read the next change in temperature, "Dropped 3°." **Ask:** *Since we don't have 3 blocks to remove, what should we do?* Instruct students to add zero pairs until they have 3 unit blocks in the positive section so they will have enough blocks to subtract from. **Say:** *You will need to add two zero pairs and then take away 3 blocks from the positive section, leaving 2 blocks on the negative section.* **Ask:** *What was the 10 A.M. reading?*

3. Have students read the next change in temperature, "Dropped 1°." **Ask:** *Since we don't have enough blocks to remove from the positive section (to subtract a positive 1), what should we do?* **Say:** *You will need to add one zero pair. Now take away 1 block from the positive side.* **Ask:** *What was the 11 A.M. reading?*

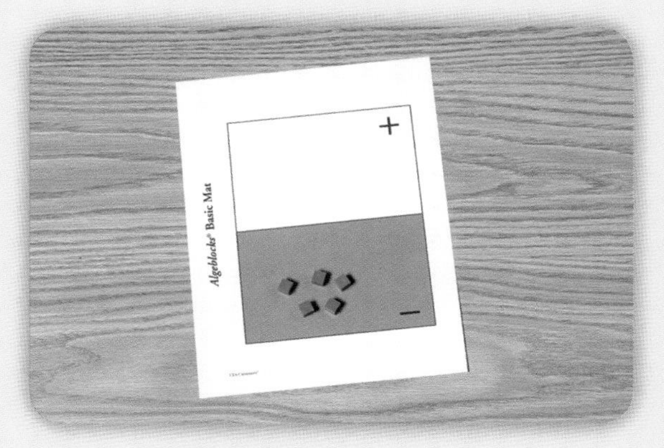

4. Say: *Now, represent the final temperature change of "Dropped 2°."* **Ask:** *What was the noon reading?*

Use Algeblocks unit blocks and a Basic Mat to model the integer subtraction sentence. Make zero pairs. Write the difference. Explain your work.

(Check students' work.)

1. $4 - (-5) =$ 9

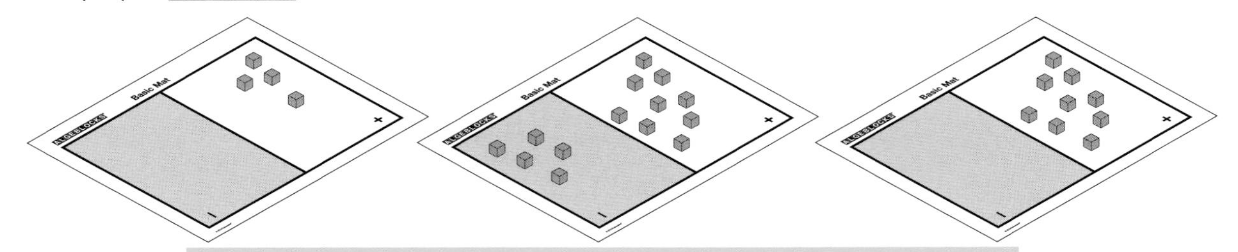

Place 4 unit blocks on the positive side of the mat. Because there are not 5 unit blocks to remove from the negative side of the mat, add 5 zero pairs to the mat. Now take away 5 unit blocks from the negative side. That leaves 9 unit blocks on the positive side of the mat.

Using Algeblocks unit blocks and a Basic Mat, model each subtraction sentence. Sketch the model. Make zero pairs. Write the difference.

2. $-6 - 4 =$ −10

3. $-9 - 7 =$ −16

Find each difference.

4. $-3 - (-1) =$ −2

5. $13 - (-7) =$ 20

6. $8 - (-12) =$ 20

7. $-5 - 11 =$ −16

8. $-1 - 6 =$ −7

9. $9 - (-8) =$ 17

10. $14 - (-16) =$ 30

11. $-15 - (-15) =$ 0

Answer Key

Challenge! For the following subtraction problems, which ones require you to place additional unit blocks that equal zero pairs so that you can take away the number being subtracted? Explain.

7 – 1   –7 – 1   7 – (–1)   –7 – (–1)

Challenge: (Sample) –7 – 1 and 7 – (–1)

Objective

Multiply integers.

Common Core State Standards

- **7.NS.2a** Understand that multiplication is extended from fractions to rational numbers by requiring that operations continue to satisfy the properties of operations, particularly the distributive property, leading to products such as (–1)(–1) = 1 and the rules for multiplying signed numbers. Interpret products of rational numbers by describing real-world contexts.

The Number System

Multiply Integers I

Students have developed the meaning of multiplication of whole numbers by using representations such as equal-size groups, arrays, area models, and equal jumps on a number line. Some of these representations also work for multiplication with negative numbers. Understanding multiplication of integers prepares students for division of integers.

Try It! *Perform the Try It! activity on the next page.*

Talk About It

Discuss the Try It! activity.

- **Say:** *When Ryan takes $5 out of his savings account, the integer –5 is used to describe the change in the amount of money in the account.* **Ask:** *When Ryan donates $5 to the food bank, what integer describes the change in the amount of money the food bank has?*
- **Say:** *The multiplication in this problem is 3 × (–5). Compare this with 3 × 5.* **Ask:** *How are they the same? How are they different?*

Solve It

Reread the problem with students. The amount of money in Ryan's savings account decreases each Friday, so a negative number (–5) is used to represent the change. To show the change in Ryan's account after 3 Fridays, students model the equation 3 × (–5) = –15. Have students explain the model.

More Ideas

For other ways to teach about multiplying integers—

- Have students use yellow and red Centimeter Cubes to model this and similar problems.
- Summarize the rules for multiplying integers.
  (1) The product of two positive integers is positive.
  (2) The product of two negative integers is positive.
  (3) The product of a positive integer and a negative integer is negative.
- Using Two-Color Counters, have students model each rule. To model the product of two negative numbers, guide students to use repeated subtraction. To subtract groups of negative quantities from zero, first add red-yellow pairs. Then take away the red counters as appropriate.

Formative Assessment

Have students try the following problem.

In a computer game, you can win a maximum of 50 points and lose a maximum of 25 points in each round. What is the lowest possible score after three rounds?

A. –150 **B.** –75 **C.** –50 **D.** –25

Try It! 15 minutes | Pairs

Here is a problem about multiplying integers.

Ryan has a savings account. Every Friday he takes out $5 from the account and donates the money to the local food bank. What is the change in Ryan's account after three Fridays?

Introduce the problem. Then have students do the activity to solve the problem. Distribute the materials.

Materials
- Two-Color Counters (at least 20 per pair)
- BLM 5

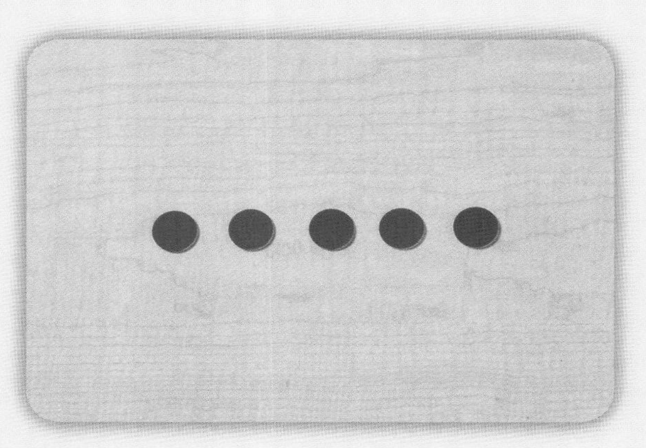

1. Say: *Let each red counter represent one dollar donated—which is one dollar less in Ryan's savings account, or –1. Use counters to show the change in Ryan's account when he makes one donation.* Students display 5 red counters.

2. Say: *Now use counters to represent the change in Ryan's account after three Fridays. Organize the counters to show that there are three equal-size groups.* Students display 3 groups of red counters, with 5 in each group. **Ask:** *What amount of money is represented?*

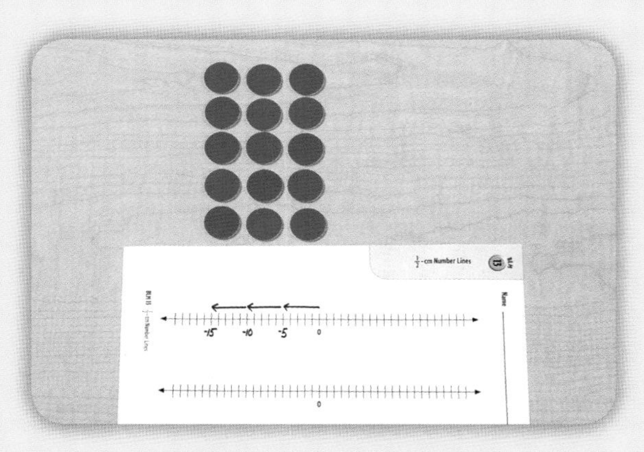

3. Say: *Now model this problem on a number line.* **Ask:** *How can you show that Ryan has $5 less in his savings account each Friday, for three Fridays?* Starting at 0, students jump 5 units left three times, ending at –15.

Look Out!

Sometimes students will be reluctant to think of multiplication as repeated addition when negative numbers are involved. Remind them that 3 × 5 is *3 groups of 5*, or 5 + 5 + 5, and that this idea applies to negative numbers, too. That is, 3 × (–5) is *three times negative five, or 3 groups of –5*, or (–5) + (–5) + (–5).

Use Two-Color Counters to model each multiplication problem. Use a number line to help. Write a number sentence for the product.

(Check students' work.)

1. 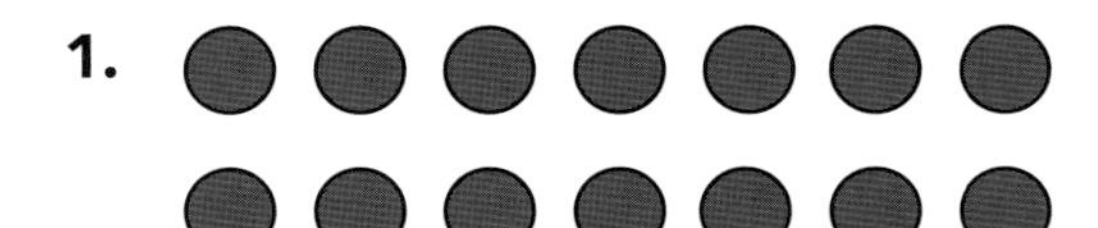

2 × (–7) = –14

2. 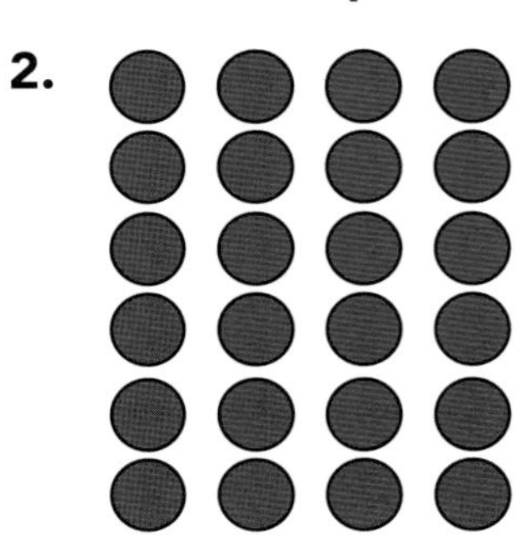

6 × (–4) = –24

Using Two-Color Counters, model each multiplication problem. Sketch the model. Write the product.

3. 7 × (–4)

–28

4. 10 × (–5)

–50

Find each product.

5. 9 × (–6) –54

6. –5 × (–7) 35

7. –4 × 11 –44

8. 9 × 7 63

9. –3 × (–1) 3

10. 8 × (–7) –56

Answer Key

Challenge! What do you notice about the product when the factors have different signs? What do you notice about the product when the factors have the same signs? Draw pictures to help.

Challenge: (Sample) When both factors have the same sign, the product is positive. When the factors have different signs, the product is negative.

Objective

Multiply integers.

Common Core State Standards

- **7.NS.2a** Understand that multiplication is extended from fractions to rational numbers by requiring that operations continue to satisfy the properties of operations, particularly the distributive property, leading to products such as (–1)(–1) = 1 and the rules for multiplying signed numbers. Interpret products of rational numbers by describing real-world contexts.

The Number System

Multiply Integers II

Everyday situations that can be represented by negative numbers, such as money owed, points lost, and descent from an elevation, help students to gain facility in computing with integers. Once they can perform addition and subtraction of integers, students can advance to the multiplication of integers.

Try It! *Perform the Try It! activity on the next page.*

Talk About It

Discuss the Try It! activity.

- **Ask:** *What does* descent *mean? Would that be represented as a positive number or a negative number? How should we represent the descent of 4 meters for the first minute on the Algeblocks® Basic Mat?*
- **Ask:** *What number represents the location of the spelunker after she has descended for 2 minutes?*
- **Ask:** *You now have 6 groups of 4 unit blocks on the mat. What number represents the level, relative to the surface, of the spelunker after 6 minutes?*

Solve It

Reread the problem with the students. Discuss the meaning of *ascent* and *descent.* Make sure that students understand how to represent each direction with integers. Ask the students to write a math sentence to represent a descent of 4 meters per minute for 6 minutes: 6 × (–4) = –24.

More Ideas

For another way to teach about multiplication of integers—

- Have students draw a vertical number line with zero labeled at the top. Students should use Centimeter Cubes to mark the descent of the spelunker for each minute. Have them mark on the number line the position of the spelunker at the end of each minute. Ask them to repeat the activity for different numbers of minutes and different rates of descent.

Formative Assessment

Have students try the following problem.

A bank charges a penalty of $35 for each check returned for insufficient funds. Suppose that a customer miscalculates his balance and writes three bad checks. Which equation expresses the effect of the penalty on his checking account?

A. (–3) × (–35) = 105 **B.** 3 × (–35) = –105

C. 3 × 35 = 105 **D.** (–3) × 35 = –105

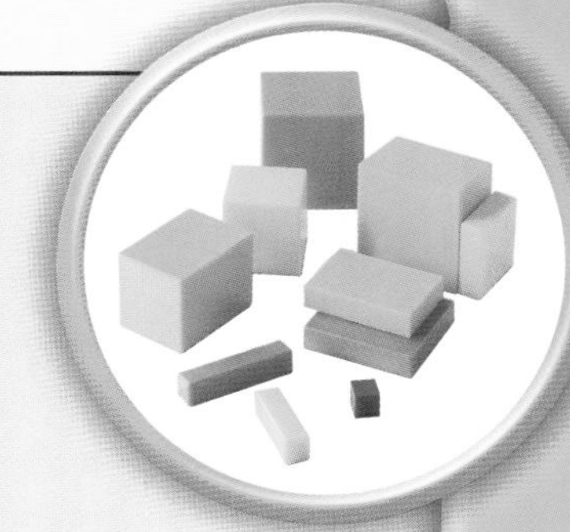

Try It! 15 minutes | Pairs

Here is a problem about multiplication of integers.

A spelunker descends into a crevasse at the rate of 4 meters per minute. What number represents her level, relative to the surface, after 6 minutes? Write a math sentence to describe this.

Introduce the problem. Then have students do the activity to solve the problem. Make sure that students understand how to represent a descent on their Algeblocks Basic Mats. Distribute the materials.

Materials
- Algeblocks® units
- BLM 4
- BLM 6
- Algeblocks Factor Track

1. Have students place 4 unit blocks on the negative section of their Algeblocks Basic Mat. **Say:** *Now you have represented the location of the spelunker after the first minute of her descent.*

2. Say: *Continue placing groups of 4 unit blocks until you have represented the spelunker's 6-minute descent.*

3. Have students complete putting 6 groups of 4 unit blocks on the left side of their mats. **Ask:** *How many unit blocks do you have on your mat? What is the depth of the spelunker after 6 minutes?* **Say:** *Write a math sentence to represent the spelunker's final location.*

4. Have students use their Algeblocks Quadrant Mats and Factor Tracks. **Say:** *Represent 6 × (–4) on the Algeblocks Quadrant Mat.* Allow time for students to do this. **Say:** *Now remove the track and read your answer. Is it the same answer you got before?*

Use Algeblocks unit blocks, a Quadrant Mat, and a Factor Track. Model each integer multiplication sentence. Find each product. (Check students' work.)

1. –2 × 3 = –6

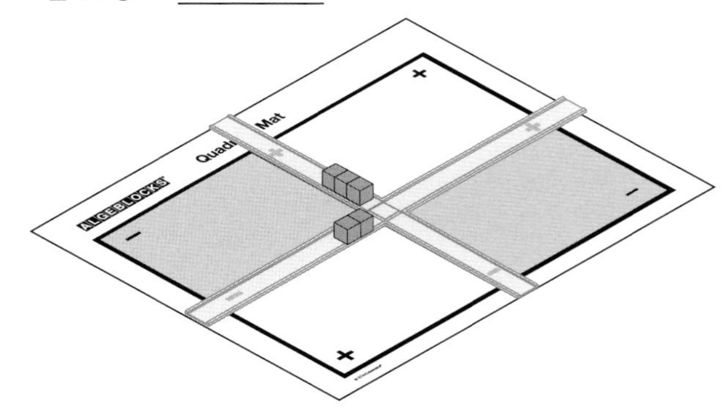

2. 3 × (–4) = –12

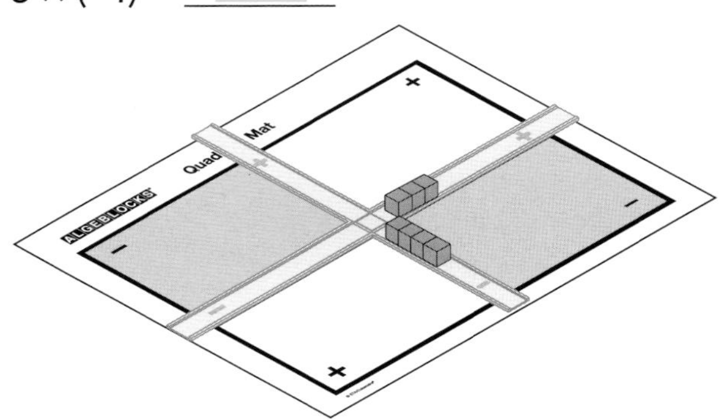

Using Algeblocks unit blocks, a Quadrant Mat, and a Factor Track, model each multiplication sentence. Sketch the model. Find the product.

3. –8 × (–2) = 16

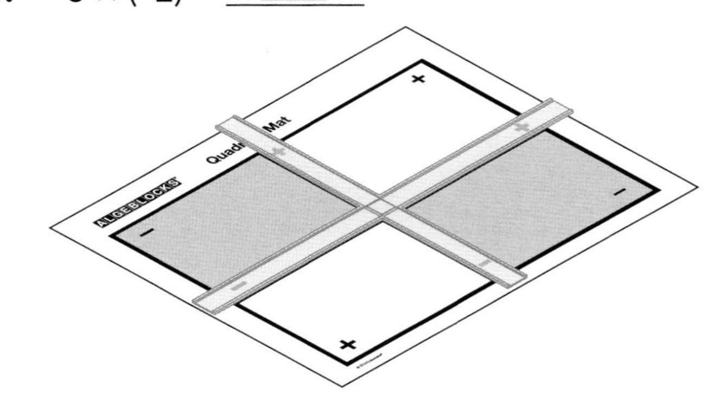

4. 5 × (–4) = –20

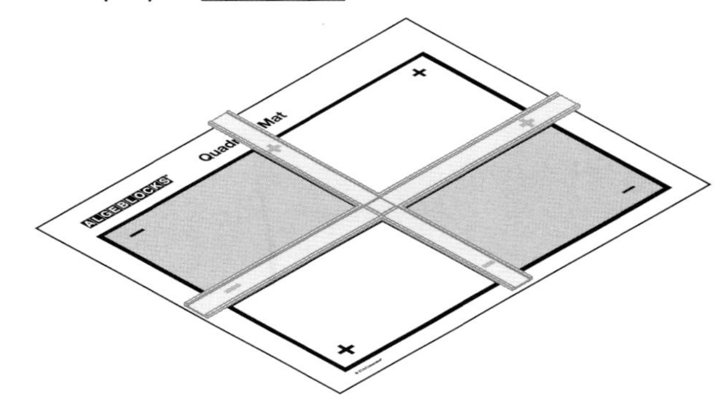

Find each product.

5. 3 × (–6) = –18

6. –7 × (–3) = 21

7. –8 × 12 = –96

8. –9 × 5 = –45

9. –5 × (–6) = 30

10. 7 × (–8) = –56

11. 11 × (–6) = –66

12. –4 × (–1) = 4

Challenge! If the product of two integers is positive, what can you conclude about the factors? Draw a picture to help.

Challenge: (Sample) If the product is positive, that means that either both factors are positive or both factors are negative.

Objective

Divide integers.

Common Core State Standards

- **7.NS.2b** Understand that integers can be divided, provided that the divisor is not zero, and every quotient of integers (with non-zero divisor) is a rational number. If p and q are integers, then $-(p/q) = (-p)/q = p/(-q)$. Interpret quotients of rational numbers by describing real-world contexts.
- **7.NS.3** Solve real-world and mathematical problems involving the four operations with rational numbers.

The Number System

Divide Integers I

Students can use what they already know about multiplying integers to divide integers. Multiplication and division are inverse operations. So, for example, to find the quotient 30 ÷ 6, students can think of the related product: 6 × ? = 30. The rules for division of positive and negative numbers are the same as those for multiplication.

Try It! *Perform the Try It! activity on the next page.*

Talk About It

Discuss the Try It! activity.

- **Say:** *The addition problem for finding the sum of the scores is –4 + 1 + (–3).* **Ask:** *Does it matter which two numbers you add first?*
- **Say:** *The division problem for finding the average score is –6 ÷ 3 = –2. Write a related multiplication problem.* Students can write either *3 × (–2) = –6,* or *(–2) × 3 = –6.*
- **Ask:** *Looking at the three scores, is it reasonable that the answer is negative rather than positive?* Have students explain their answers.

Solve It

Reread the problem with students. The average score is the sum of the scores divided by the number of scores: [–4 + 1 + (–3)] ÷ 3. The sum of the scores is –6 and –6 ÷ 3 is –2. Ken's average score is –2.

More Ideas

For other ways to teach about dividing integers—

- Have students use Centimeter Cubes to model pairs of number sentences, such as –10 ÷ 5 = –2 and 5 × (–2) = –10. Suggest that students use red cubes for negative numbers and yellow cubes for positive numbers. Also, discuss why –8 ÷ 4 is easier to model than 8 ÷ (–4).
- Summarize the rules for dividing integers, and note that they are the same as the rules for multiplying integers.
  (1) The quotient of two positive integers is positive.
  (2) The quotient of two negative integers is positive.
  (3) The quotient of a positive integer and a negative integer is negative.
- Using Two-Color Counters, have students model an equation for rules 1 and 3.

Formative Assessment

Have students try the following problem.

Find the average of –6, –4, 8, and –2.

A. –4 **B.** –1 **C.** 1 **D.** 4

Try It! 15 minutes | Pairs

Here is a problem about dividing integers.

In three rounds of golf, Ken shot scores of –4, +1, and –3. What was his average score?

Introduce the problem. Then have students do the activity to solve the problem. Distribute the materials. Remind students that the average is the sum of the scores divided by the number of scores.

Materials

- Two-Color Counters (at least 20 per pair)

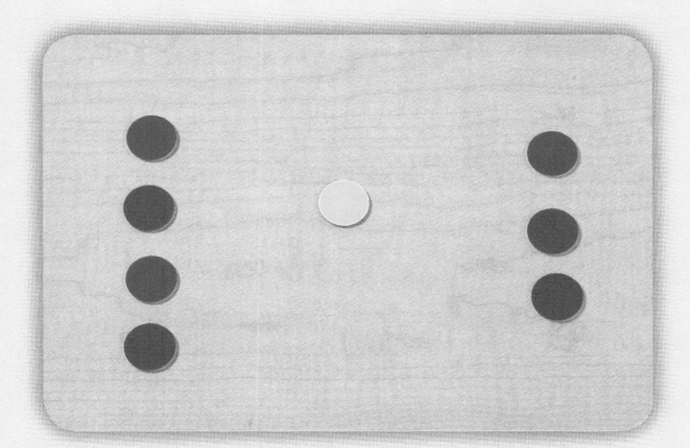

1. Say: *Each red counter represents negative one, and each yellow counter represents positive one. Use counters to show Ken's three scores.* Students place 4 red counters together, 1 yellow counter by itself, and 3 red counters together.

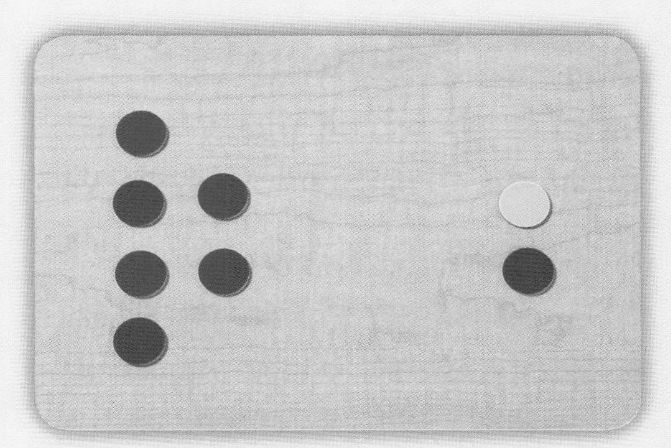

2. Say: *To find the average score, you first need to add the three scores. Use the counters to find the sum.* Students pair one yellow counter with a red counter to equal zero, and move the pair aside. There are 6 red counters left. The sum is –6.

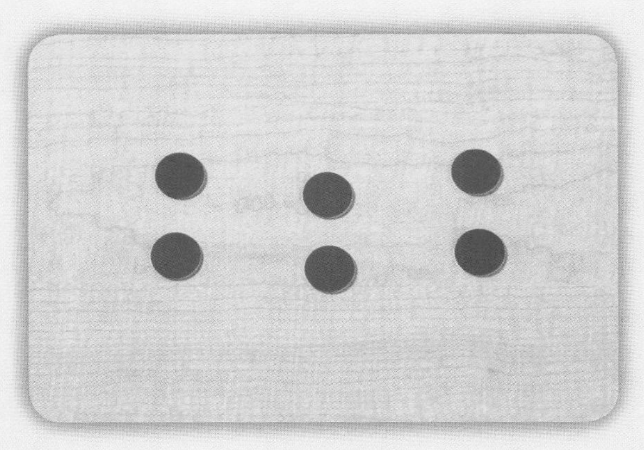

3. Say: *Now divide the sum by 3, since there are 3 scores. Divide the counters that represent the sum into 3 equal groups.* **Ask:** *How many counters are in each group?* Students arrange the 6 red counters into 3 groups, with 2 red counters in each group.

Make sure students understand that the average does not have to be one of the scores. To calculate the average score, they must add all the scores and divide by the number of scores, even though the scores include both positive and negative numbers. Note that the given scores, written in order from least to greatest, are –4, –3, and +1. It makes sense that the average, which is –2, lies between the least score, –4, and the greatest score, +1.

Use Two-Color Counters to model each division problem. Write a number sentence for the quotient.

(Check students' work.)

1.

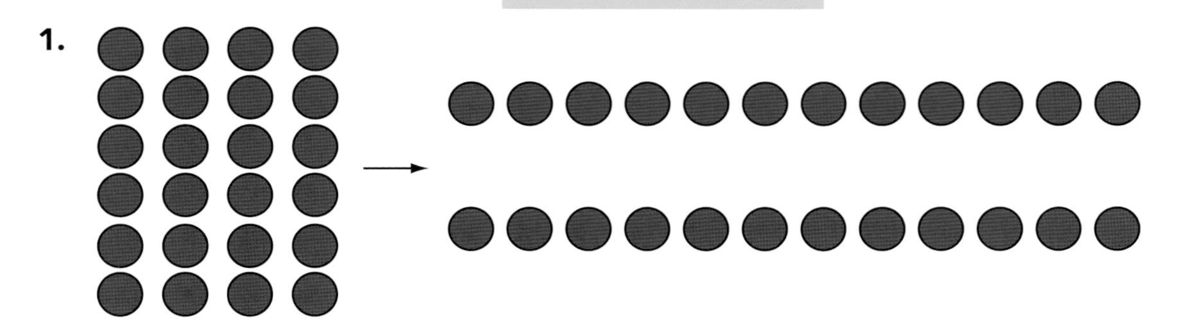

−24 ÷ 2 = −12

Using Two-Color Counters, model each division problem. Sketch the model. Find the quotient.

2. −35 ÷ 7

−5

3. −81 ÷ 9

−9

Find each quotient.

4. 49 ÷ (−7) = −7

5. −45 ÷ (−5) = 9

6. −42 ÷ 7 = −6

7. 9 ÷ (−3) = −3

8. −30 ÷ (−6) = 5

9. 28 ÷ (−7) = −4

Challenge! How do the rules for dividing integers differ from the rules for multiplying integers? Draw pictures to help.

Challenge: (Sample) The rules for multiplying and dividing integers are the same.

Objective

Divide integers.

Common Core State Standards

- **7.NS.2b** Understand that integers can be divided, provided that the divisor is not zero, and every quotient of integers (with non-zero divisor) is a rational number. If p and q are integers, then $-(p/q) = (-p)/q = p/(-q)$. Interpret quotients of rational numbers by describing real-world contexts.

The Number System

Divide Integers II

To extend their understanding of integers to division of integers, students need to draw on their understanding of multiplication and division as inverse operations. Using the area-array model for multiplication will help students use the Algeblocks® Quadrant Mat effectively.

Try It! *Perform the Try It! activity on the next page.*

Talk About It

Discuss the Try It! activity.

- **Ask:** *What integer would represent a $14 loss? Where should we place the 14 unit blocks?*
- **Ask:** *What are the factors of 14?*
- **Ask:** *Where should we place the divisor of 2?*
- **Ask:** *What would the other factor, the quotient, be?*

Solve It

Reread the problem with the students. Help them represent the financial loss on their Algeblocks Quadrant Mats. Suggest that they make groups of two unit blocks, since there are two business partners. The number of groups represents the number of dollars each partner lost. Have students write a math sentence to represent the situation.

More Ideas

For another way to teach about division of integers—

- Challenge students to write a number sentence and represent it using the Algeblocks Quadrant Mat and Factor Track for each of the following situations:

  a. a positive number divided by a positive number;

  b. a positive number divided by a negative number;

  c. a negative number divided by a positive number;

  d. a negative number divided by a negative number.

Formative Assessment

Have students try the following problem.

The water level in a swollen river falls 30 cm in 5 hours. What is the average change in its level each hour?

A. 15 cm per hour

B. 6 cm per hour

C. –15 cm per hour

D. –6 cm per hour

Try It! 15 minutes | Pairs

Here is a problem about dividing a negative integer.

Jessica and Taylor decided to start a dog-walking business for the summer. After paying for lessons in dog handling and some basic equipment, they began signing up customers. At the end of the summer, they added up the money they had earned, subtracted their expenses, and found that they had actually lost $14! What is the financial outcome of this business venture for each of the two partners?

Introduce the problem. Then have students do the activity to solve the problem. Distribute the materials.

Materials
- Algeblocks® units
- BLM 6
- Algeblocks Factor Track

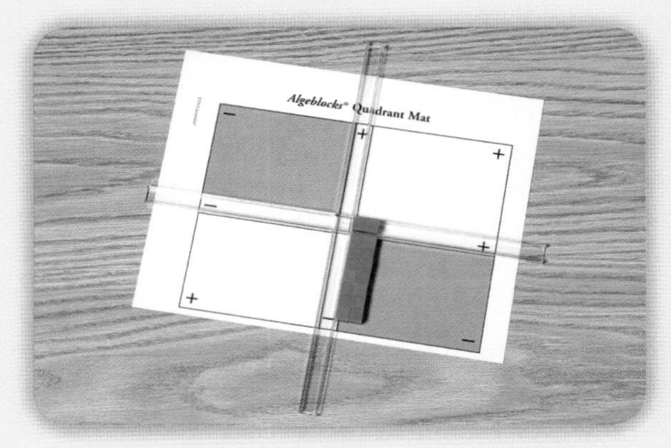

1. Have students represent on their Algeblocks Quadrant Mat the money lost in the business venture. **Ask:** *What are the factors of 14?* **Ask:** *How many business partners were there?* Have students place 2 unit blocks on the positive side of the horizontal bar of the Factor Track. **Say:** *Now, arrange 14 unit blocks in Quadrant IV in a rectangle with 2 unit blocks on one side.*

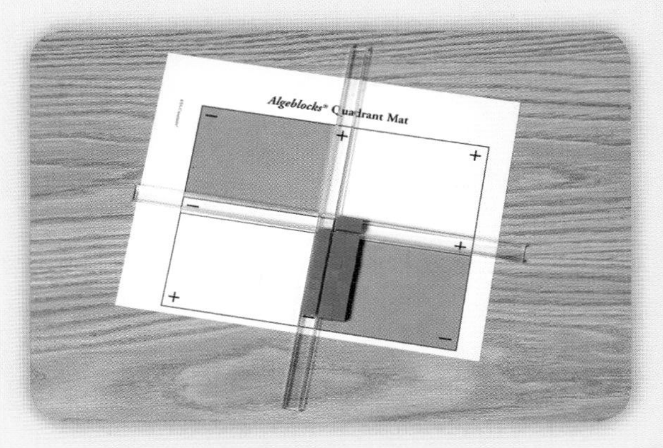

2. Ask: *How many groups of 2 do you have?* Have students place 7 unit blocks on the negative part of the vertical bar of the Factor Track. **Say:** *Now you have shown that each partner lost $7.*

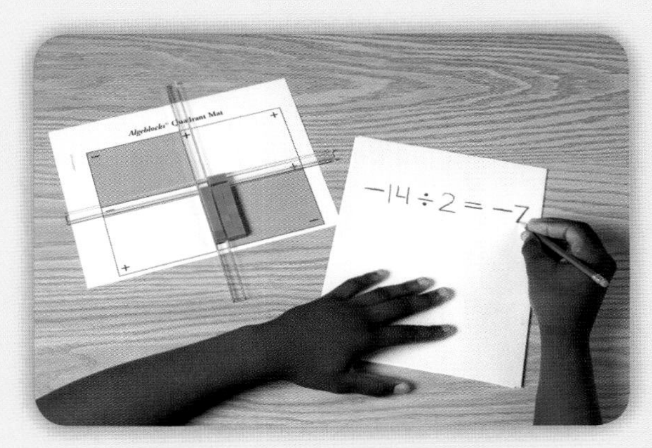

3. Have students write a number sentence to represent each partner's loss.

⚠ Look Out!

Watch for students who are unsure of where to place the 14 unit blocks on the Quadrant Mat. Since the business partners lost $14, the 14 blocks should go in either of the gray areas. (If students choose to put the blocks in Quadrant II, they should take care to place the two blocks, representing two partners, on the positive area of the Factor Track.) In this example, the 14 blocks were placed in the lower right and the other blocks were placed accordingly.

Use Algeblocks unit blocks, a Quadrant Mat, and a Factor Track. Model each integer division sentence. Find each quotient. (Check students' work.)

1. 15 ÷ (–3) = –5

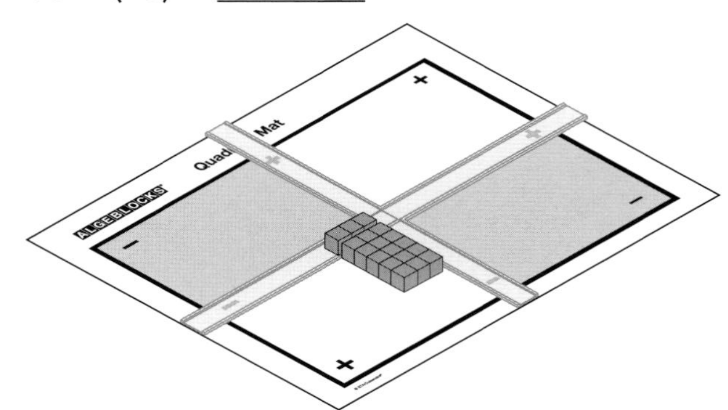

2. –20 ÷ (5) = –4

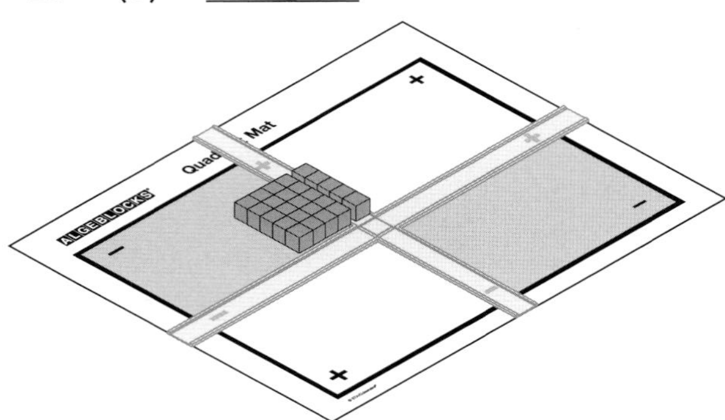

Using Algeblocks unit blocks, a Quadrant Mat, and a Factor Track, model each division sentence. Sketch the model. Find each quotient.

3. –28 ÷ (–7) = 4

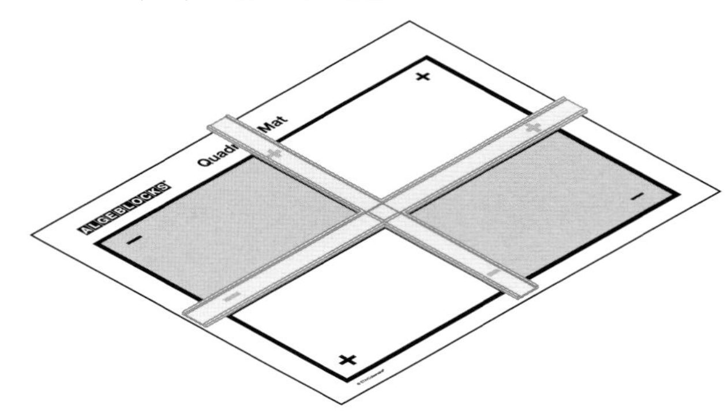

4. 45 ÷ (5) = 9

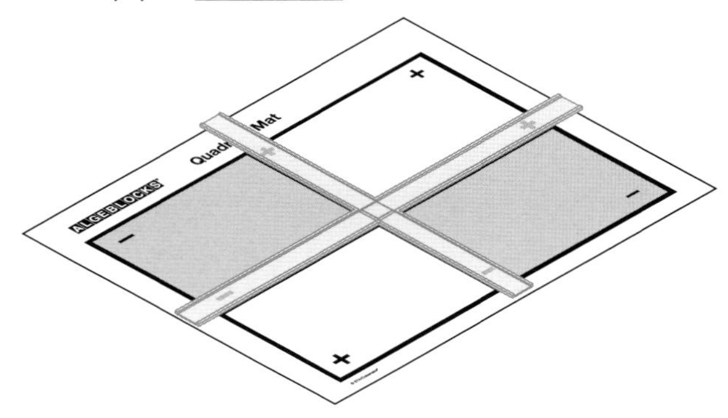

Find each quotient.

5. 36 ÷ (–6) = –6

6. –27 ÷ (–3) = 9

7. –18 ÷ 3 = –6

8. –49 ÷ 7 = –7

9. –35 ÷ 5 = –7

10. 12 ÷ (–2) = –6

11. –24 ÷ (–4) = 6

12. –5 ÷ (–5) = 1

Answer Key

Challenge! How do the rules for adding and subtracting integers differ from the rules for multiplying and dividing integers?

Challenge: (Sample) When adding or subtracting, the sum or difference depends on the magnitude of the numbers. A positive plus a negative can be either positive or negative. When mutliplying or dividing, the product or quotient of a negative and a positve will always be negative.

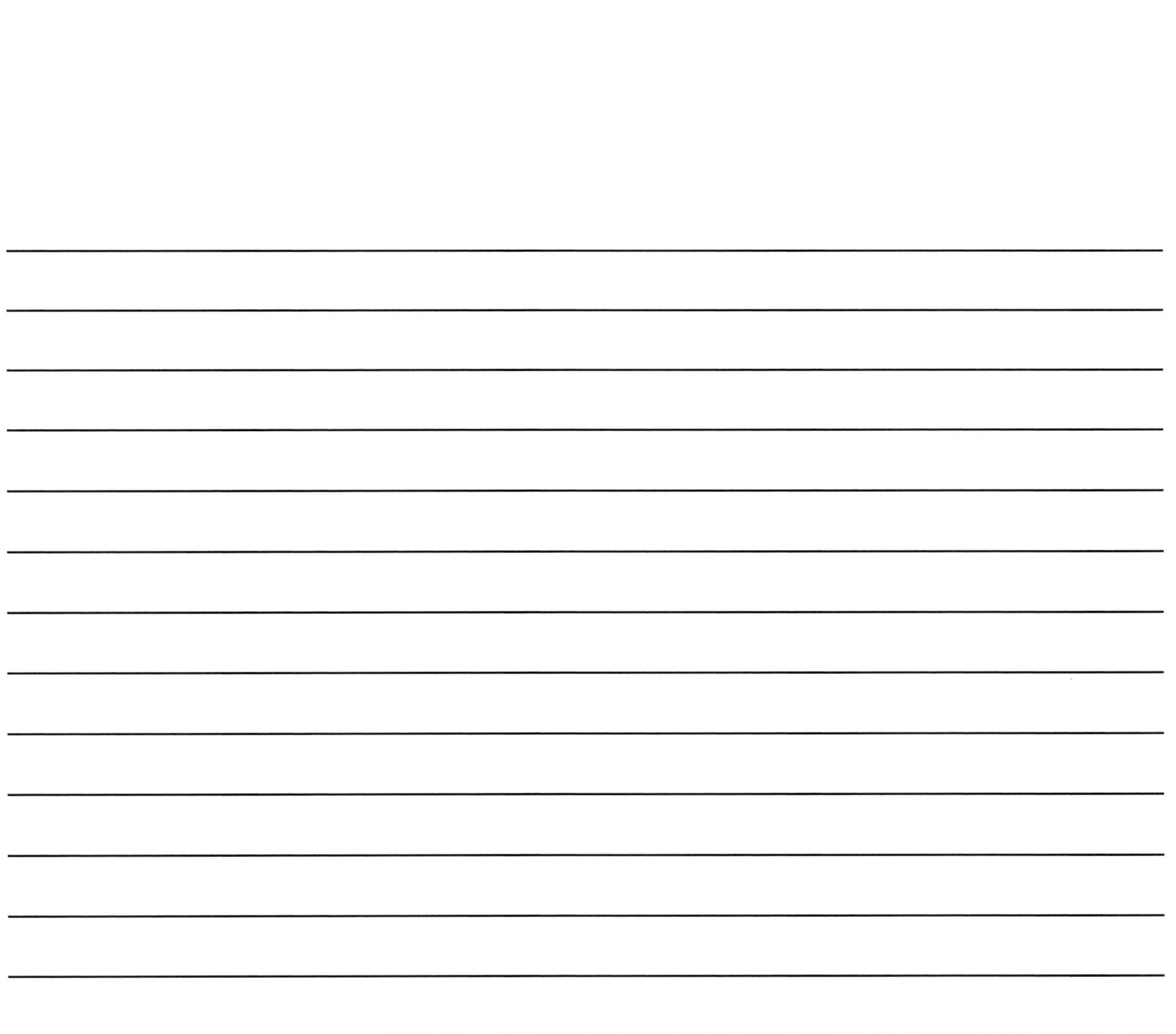

Expressions and Equations

In **Expressions and Equations**, students continue to develop their understanding of the order of operations with rational numbers. The **order of operations** is a set of rules for determining the order in which the operations in an expression are performed. A **rational number** is a number that can be expressed as a fraction, or ratio of two integers, where the denominator is not zero. Fractions, decimals, integers, and percents are all rational numbers. Students also continue to develop understanding of dependent and independent variables (letters that represent numbers), and use them to represent quantities in real-world or mathematical problems.

Additionally, students use the properties of operations to add, subtract, factor, and expand linear expressions with rational coefficients. **Linear expressions** are expressions in which no variable is raised to a power other than 1. For example, $3x + 6$ is a linear expression. A coefficient is a number in front of a variable that indicates the quantity of the variable. For example, 3 is a coefficient in the expression $3x + 6$, and 2 and 5 are coefficients in the equation $2x + 5y + 3 = 29$.

At this level, students expand their mathematical versatility by using equivalent expressions to solve problems. Students learn that changing the form of an expression can shed light on its role in a problem and on how the quantities in the problem are related. Students assess the reasonableness of answers by using mental computation and estimation strategies.

The Grade 7 Common Core State Standards for Expressions and Equations specify that students should–

- Use properties of operations to generate equivalent expressions.
- Solve real-life and mathematical problems using numerical and algebraic expressions and equations.

The following hands-on activities enable teachers to help students understand the principles of expressions and equations. The activities will help students use expressions and equations to flexibly tackle algebraic thinking and reasoning tasks.

Expressions and Equations

Contents

Objective

Write a number as a mixed number, a decimal, and a percent greater than 100%.

Common Core State Standards

- **7.EE.3** Solve multi-step real-life and mathematical problems posed with positive and negative rational numbers in any form (whole numbers, fractions, and decimals), using tools strategically. Apply properties of operations to calculate with numbers in any form; convert between forms as appropriate; and assess the reasonableness of answers using mental computation and estimation strategies. *For example: If a woman making $25 an hour gets a 10% raise, she will make an additional 1/10 of her salary an hour, or $2.50, for a new salary of $27.50. If you want to place a towel bar 9 3/4 inches long in the center of a door that is 27 1/2 inches wide, you will need to place the bar about 9 inches from each edge; this estimate can be used as a check on the exact computation.*

Expressions and Equations

Mixed Numbers, Decimals, and Percents Greater than 100%

Students expand their experiences with different number representations by looking at mixed numbers and their equivalent decimals and percents. Using models helps students increase their flexibility with these numbers. As their number fluency increases, students begin to differentiate between situations in which one representation may be more suitable than another.

Try It! *Perform the Try It! activity on the next page.*

Talk About It

Discuss the Try It! activity.

- **Ask:** *Could you compare the numbers in the same way if each person had used a different-size carton to hold the balls? Why or why not?*
- Have students explain how to write $1\frac{1}{12}$ as a percent.
- Have students describe situations in which it might be better to choose one form over another to represent a number. For example, when using money, a decimal is the more accepted number representation.

Solve It

Reread the problem with students. Have students compare the fractional portions of their models to answer the question.

More Ideas

For other ways to teach equivalency among mixed numbers, decimals, and percents greater than 100%—

- Have students use Fraction Tower® Equivalency Cubes to model the problem. Have students see that one whole equals 1, 1.0, and 100%, regardless of the number of equal-size parts in the whole. Have students show amounts greater than one whole as they explore how different values can be combined to show mixed numbers, decimals greater than one, and percents greater than 100%.
- Have students cover Base Ten Flats with Deluxe Rainbow Fraction® Squares to model amounts greater than one whole, then write each amount in the three forms using their knowledge of percents, decimals, and fractions. Some students may find it helpful to shade 10 × 10 grids (BLM 13) to show each amount and to help them to rewrite the amounts in each of the three forms.

Formative Assessment

Have students try the following problem.

Which equation is true?

A. $1\frac{7}{10}$ = 1,710% **B.** 20.0 = 200% **C.** $1\frac{1}{5}$ = 120% **D.** 2.45 = 2.45%

Try It! 20 minutes | Groups of 4

Here is a problem about mixed numbers, decimals, and percents.

Randi, Nick, and LaKeisha collected golf balls from a pond on the golf course and put them into cartons. Each carton holds the same number of balls. Randi filled $1\frac{3}{8}$ cartons, Nick filled 1.5 cartons, and LaKeisha filled 175% of a carton. Who found the most golf balls?

Introduce the problem. Then have students do the activity to solve the problem. Distribute the materials. Review the ways that one whole can be represented: 1, 1.0, and 100%. In the problem, a whole is one carton of golf balls.

Materials

- Deluxe Rainbow Fraction® Squares (1 set per group)
- paper (3 sheets per group)

1. Write $1 + \frac{3}{8} = 1\frac{3}{8}$ on the board. **Say:** *A mixed number is the sum of its whole-number part and its fraction part. Use fraction pieces to show $1\frac{3}{8}$.* Have students rename the fraction part as a decimal and as a percent, use the correct names for the whole, and write the corresponding equations.

2. Have students show 1.5. **Ask:** *How can we write an equation to add the whole-number part and the decimal part of this number?* Write *1.0 + 0.5 = 1.5* on the board. **Ask:** *How can we write five-tenths as a fraction in simplest form and as a percent?* Have students write the three equations shown by the model.

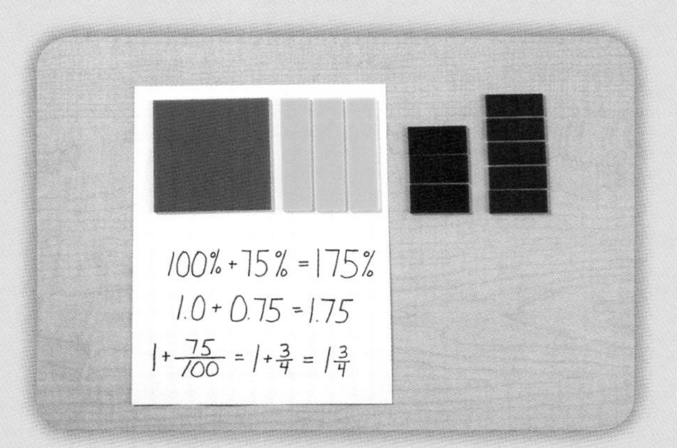

3. Write *100% + ? = 175%* on the board. **Ask:** *What percent can replace the question mark?* Discuss how to use fraction pieces to show 75%. Have students model the equation and write three equations shown by the model.

4. Have students recognize that the red square represents 1 and is the same in each case. Students can therefore answer the question by comparing the fractional portions of their models.

Use Fraction Squares to model each mixed number. Write a number sentence for the mixed number model. Write number sentences for the decimal and for the percent.

(Check students' work.)

1.

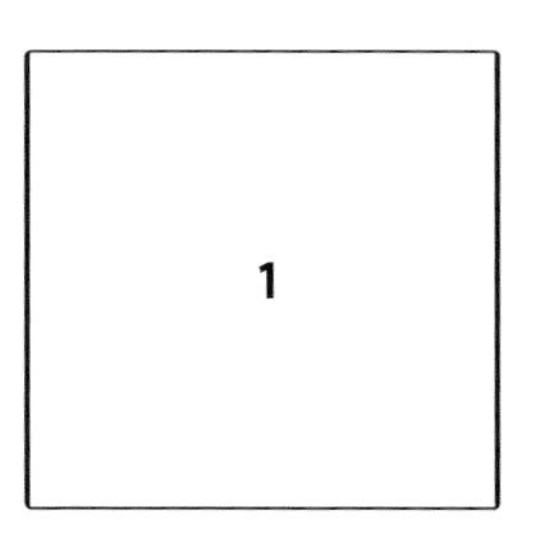

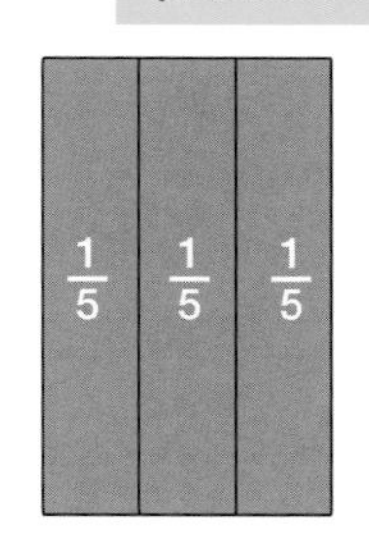

mixed number: $1 + \frac{3}{5} = 1\frac{3}{5}$

decimal: 1 + 0.6 = 1.6

percent: 100% + 60% = 160%

Using Fraction Squares, model each number. Write number sentences for the mixed number, decimal, and percent.

2.

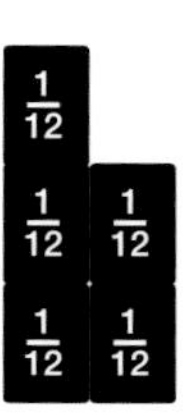

mixed number: $1 + \frac{5}{12} = 1\frac{5}{12}$

decimal: $1 + 0.41\overline{6} = 1.41\overline{6}$

percent: $100\% + 41.\overline{6}\% = 141.\overline{6}\%$

3.

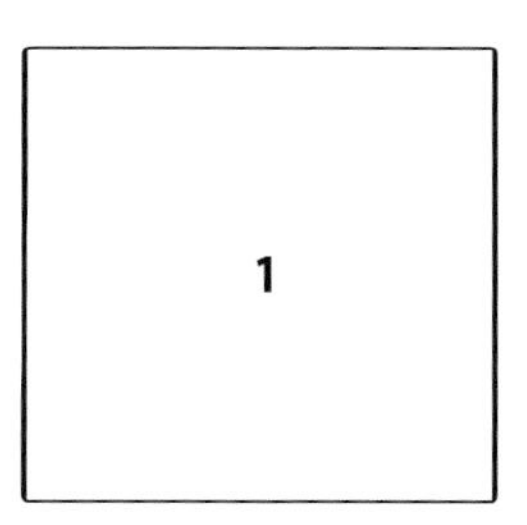

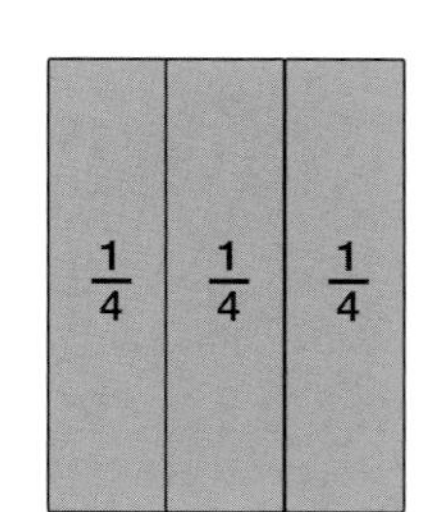

mixed number: $1 + \frac{3}{4} = 1\frac{3}{4}$

decimal: 1 + 0.75 =1.75

percent: 100% + 75% = 175%

Write each mixed number as a decimal and as a percent.

4. $1\frac{1}{3}$

$1.\overline{3}$

$133.\overline{3}\%$

5. $1\frac{4}{5}$

1.8

180%

6. $2\frac{1}{4}$

2.25

225%

7. $1\frac{5}{6}$

$1.8\overline{3}$

$183.\overline{3}\%$

8. $2\frac{2}{3}$

$2.\overline{6}$

$266.\overline{6}\%$

9. $1\frac{7}{8}$

1.875

187.5%

Answer Key

Challenge! Compare the mixed numbers in Questions 1, 2, and 3. Write the numbers as decimals from greatest to least. Explain how you compared the numbers.

Challenge: (Sample) 1.75, 1.6, $1.41\overline{6}$; I looked at the whole number. Because all were 1, I looked at the tenths place. 7 > 6 > 4

Objective

Convert fractions into decimals and percentages.

Common Core State Standards

- **7.EE.3** Solve multi-step real-life and mathematical problems posed with positive and negative rational numbers in any form (whole numbers, fractions, and decimals), using tools strategically. Apply properties of operations to calculate with numbers in any form; convert between forms as appropriate; and assess the reasonableness of answers using mental computation and estimation strategies. *For example: If a woman making $25 an hour gets a 10% raise, she will make an additional 1/10 of her salary an hour, or $2.50, for a new salary of $27.50. If you want to place a towel bar 9 3/4 inches long in the center of a door that is 27 1/2 inches wide, you will need to place the bar about 9 inches from each edge; this estimate can be used as a check on the exact computation.*

Expressions and Equations

Converting Fractions, Decimals, and Percentages

In previous grades, students learned that the set of rational numbers consists of all numbers of the form $\frac{p}{q}$, where p and q are integers and $q \neq 0$. Students should also be familiar with reducing fractions. In this lesson, students will use their previous knowledge of fractions to convert fractions to both decimals and percentages.

Try It! *Perform the Try It! activity on the next page.*

Talk About It

Discuss the Try It! activity.

- **Ask:** *What is a percentage?* Elicit from students that it is the ratio of some number to 100. **Ask:** *How do you change a fraction to a percentage?*
- **Ask:** *To change a decimal to a percentage, how many places should you move the decimal point?* Then have students explain how to change a percentage to a decimal.

Solve It

Reread the problem with students. Have students generate three reduced fractions from the story problem. Have them convert each fraction to a decimal and then each decimal to a percentage.

More Ideas

For other ways to teach about fractions, decimals, and percentages—

- Give students about 200 Centimeter Cubes in five colors. Have students randomly select 100 cubes, note the number of cubes of each color, express each color as a fraction of 100, reduce all fractions that are not in lowest terms, and convert each fraction to a decimal and to a percentage.
- Have students use Fraction Tower® Equivalency Cubes to solve similar problems. Any combination of cubes that can be stacked to the same height as the red cube will be equal to 100% $\left(\text{e.g., } \frac{1}{8}, \frac{1}{8}, \frac{1}{4}, \frac{1}{6}, \text{ and } \frac{1}{3}\right)$. The other sides of the cubes show decimal and percent equivalents.

Formative Assessment

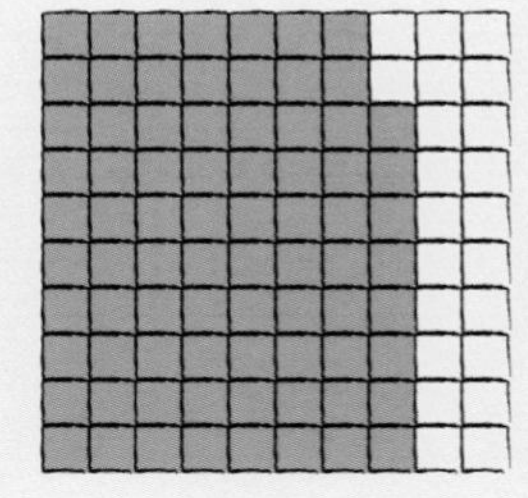

Have students try the following problem.

Which of the following sets of fractions, decimals, and percentages is represented by the shaded area of the 10 × 10 grid shown here?

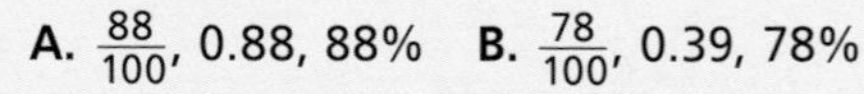

A. $\frac{88}{100}$, 0.88, 88% **B.** $\frac{78}{100}$, 0.39, 78%

C. $\frac{39}{50}$, 0.78, 78% **D.** $\frac{22}{25}$, 0.88, 88%

Try It! 20 minutes | Groups of 4

Here is a problem about fractions, decimals, and percentages.

A marketing company conducted a survey of one hundred 13- to 18-year-olds asking them to name their favorite type of movie. The results were as follows:

Action: 26	*Romance: 15*
Science Fiction: 30	*Comedy: 29*

Introduce the problem. Then have students do the activity to solve the problem. Distribute the materials.

Materials

- Color Tiles (33 of each color)

1. Have students form single-color groups of color tiles to reflect the number of responses to each survey category. Use the following color key:
Action = red; Science Fiction = blue;
Romance = green; Comedy = yellow.

2. Have students form a 10-by-10 array with the tiles, grouping the colors to reflect the survey results.

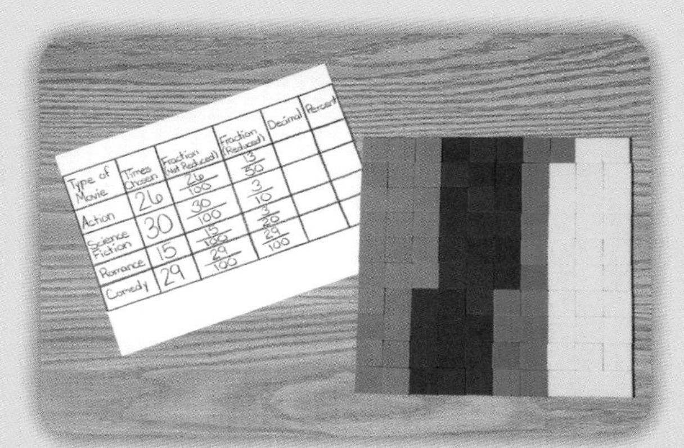

3. Have students make a table with six columns: Type of Movie, Times Chosen, Fraction (Not Reduced), Fraction (Reduced), Decimal, and Percent. Tell students to fill in the first two columns based on information from the problem and to fill in the fractions in the Not Reduced column by writing the number of times that type of film was chosen over 100. Finally, have students reduce the fractions.

4. Have students write the decimal form for each result by looking at the numerator of the fractions that have not been reduced. Since 100 is the denominator, the decimal point should be put in front of any two-digit number (e.g., $\frac{22}{100}$ would be written as 0.22). Students can convert the decimal to a percentage by moving the decimal point two places to the right and writing a percent sign after the numeral.

Use Color Tiles in a 10 × 10 array to model the fraction shown. Write the fraction for each color. Then write the decimal and percent for each color.

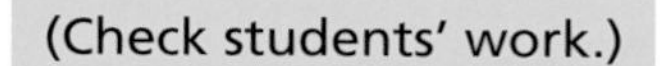

1.

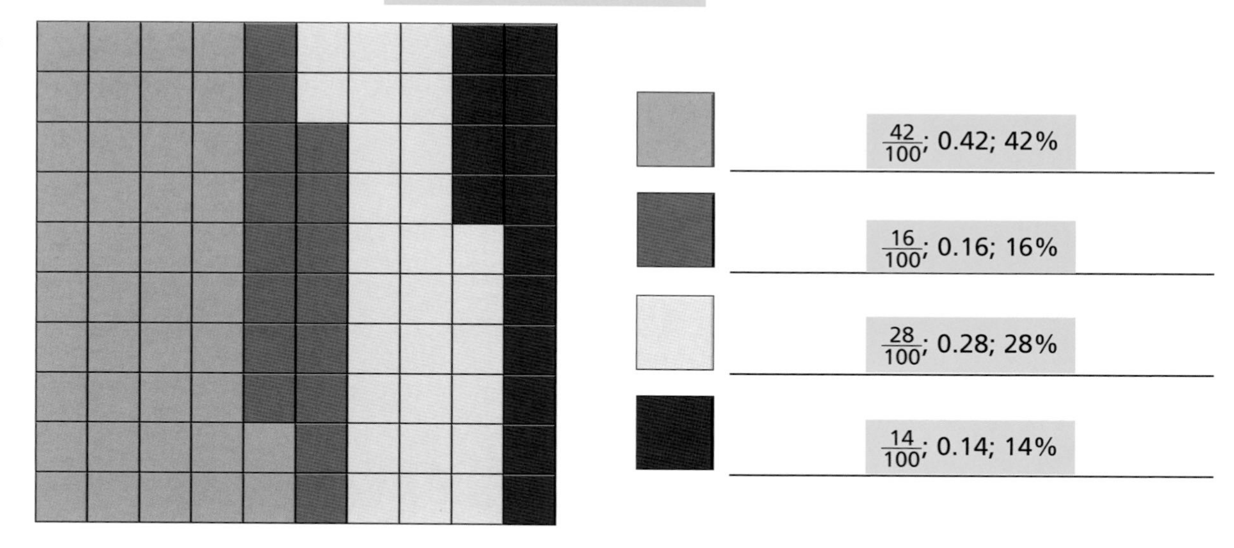

Using Color Tiles, model a 10 × 10 array for the fractions given. Sketch the model. Write the decimal and percent for each color.

2.

Red: $\frac{35}{100}$ 0.35; 35%

Blue: $\frac{12}{100}$ 0.12; 12%

Yellow: $\frac{32}{100}$ 0.32; 32%

Green: $\frac{21}{100}$ 0.21; 21%

Write each fraction as a decimal and as a percent.

3. $\frac{18}{100}$ 0.18; 18%

4. $\frac{33}{100}$ 0.33; 33%

5. $\frac{72}{100}$ 0.72; 72%

6. $\frac{25}{100}$ 0.25; 25%

7. $\frac{16}{100}$ 0.16; 16%

8. $\frac{40}{100}$ 0.40; 40%

Answer Key

Challenge! Why can you use a 10 × 10 array to convert a part of a total to a percent?

Challenge: (Sample) A 10 × 10 array has 100 units. Percent means per one hundred.

Objective

Convert a percentage to a fraction.

Common Core State Standards

- **7.EE.3** Solve multi-step real-life and mathematical problems posed with positive and negative rational numbers in any form (whole numbers, fractions, and decimals), using tools strategically. Apply properties of operations to calculate with numbers in any form; convert between forms as appropriate; and assess the reasonableness of answers using mental computation and estimation strategies. *For example: If a woman making $25 an hour gets a 10% raise, she will make an additional 1/10 of her salary an hour, or $2.50, for a new salary of $27.50. If you want to place a towel bar 9 3/4 inches long in the center of a door that is 27 1/2 inches wide, you will need to place the bar about 9 inches from each edge; this estimate can be used as a check on the exact computation.*

Expressions and Equations

Fraction, Decimal, and Percentage Combinations that Equal 1

Previously, students learned to convert fractions to percentages. They divided the numerator by the denominator, multiplied the quotient by 100, and added a percent sign. In this lesson, students will work these steps backward to convert percentages to fractions.

Try It! *Perform the Try It! activity on the next page.*

Talk About It

Discuss the Try It! activity.

- **Ask:** *What fraction of the circle does each section represent?*
- **Ask:** *What is the decimal equivalent of each of the fractions?*
- Have students explain how to convert a percentage to a decimal and then a decimal to a fraction without using the Rainbow Fraction Circle Rings.

Solve It

Reread the problem with students. Have them convert the percentages to decimals. Then have students use the Rainbow Fraction® Circle Rings to determine the equivalent fractions in order to solve the story problem.

More Ideas

For another way to teach about fractions, decimals, and percentages—

- Students can repeat this activity using Fraction Tower® Equivalency Cubes to represent the various types of glass found in Roberto's artwork.

Formative Assessment

Have students try the following problem.

Which of the following sets of a fraction, decimal, and percentage represents the missing section of the circle graph shown here?

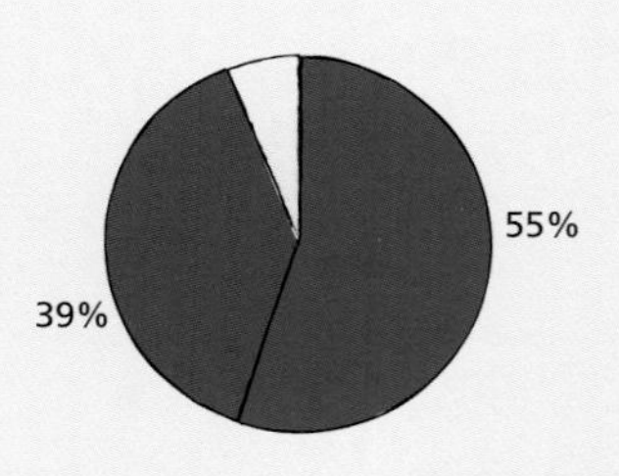

A. $\frac{6}{100}$, .6, 6% **B.** $\frac{3}{50}$, .06, 6%

C. $\frac{8}{50}$, .08, 16% **D.** $\frac{16}{100}$, .16, 16%

Try It! 40 minutes | Groups of 4

Here is a problem about fractions, decimals, and percentages.

Roberto is designing a circular glass work of art for his Fine Arts class. He would like the circle to be divided into four sections, each with a different type of glass. The sections should be 20% beveled glass, 16.$\overline{6}$% swirled glass, 30% bubbled glass, and 33.$\overline{3}$% wrinkled glass. What fraction of the circle will each texture be?

Introduce the problem. Then have students do the activity to solve the problem. Distribute the materials.

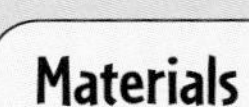

Materials

- Deluxe Rainbow Fraction® Circles
- Rainbow Fraction Circle Rings

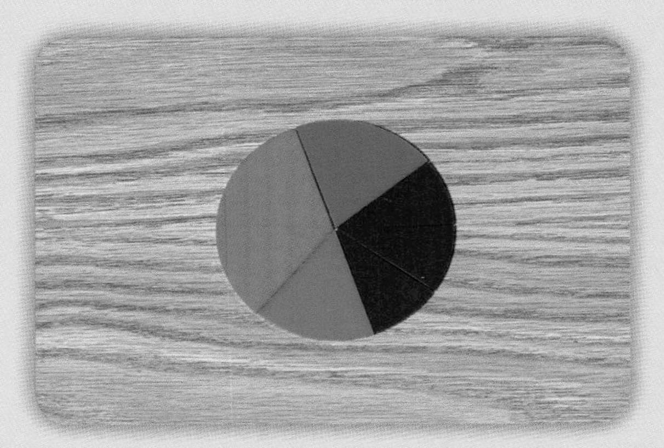

1. Have students use the Fraction Circles to represent the percentages for each type of glass. The Percent Ring will help students choose the appropriate pieces. **Ask:** *What Fraction Circle piece will you use to represent the beveled glass section? The wrinkled glass section? How will you represent the bubbled glass?*

2. Now have students convert the percentages to decimals. **Ask:** *Which way should you move the decimal point, left or right? How many places should you move it?*

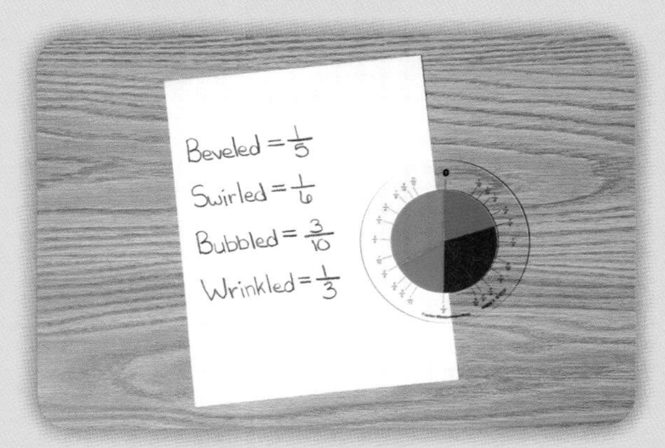

3. Have students use the Fraction Circle Rings to determine which fraction each decimal represents. Have students record their results on a sheet of paper.

⚠ Look Out!

Some students may add the 16.$\overline{6}$% and the 33.$\overline{3}$% and write the sum as 49.$\overline{9}$%. Explain to them that the repeating bar ($\overline{\ }$) indicates that the last digit repeats endlessly and that the number is therefore an approximation. These percentages represent $\frac{1}{6}$ and $\frac{1}{3}$, which add up to $\frac{1}{2}$. It is understood that in these situations, the .$\overline{6}$ and the .$\overline{3}$ equal 1 when added, and so 16.$\overline{6}$% plus 33.$\overline{3}$% equals 50.0%.

Use Fraction Circles and Fraction Circle Rings to model each percentage. Write the percent as a fraction.

(Check students' work.)

1. 25% $\frac{1}{4}$

33.$\overline{3}$% $\frac{1}{3}$

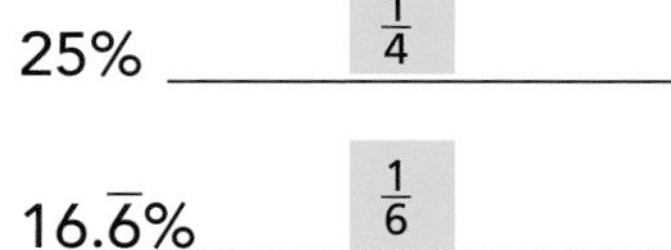

25% $\frac{1}{4}$

16.$\overline{6}$% $\frac{1}{6}$

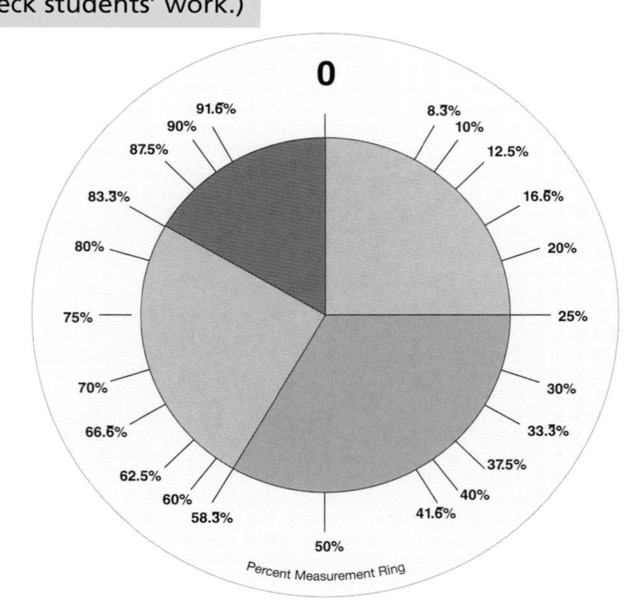

Using Fraction Circles and Fraction Circle Rings, model each percentage. Sketch the model. Write the percent as a fraction.

2. 20% $\frac{1}{5}$

12.5% $\frac{1}{8}$

37.5% $\frac{3}{8}$

30% $\frac{3}{10}$

Write each percent as a fraction.

3. 80% $\frac{4}{5}$

4. 62.5% $\frac{5}{8}$

5. 16.$\overline{6}$% $\frac{1}{6}$

6. 87.5% $\frac{7}{8}$

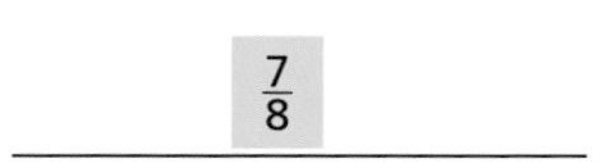

7. 41.$\overline{6}$% $\frac{5}{12}$

8. 75% $\frac{3}{4}$

Answer Key

Challenge! What does the word *percent* mean? Explain how to get the numerator and decimal of a fraction equivalent to a given percent.

Challenge: (Sample) *Percent* means per 100. The denominator of the fraction equivalent to a percent is 100. The numerator is the number of the percent written without a percent symbol.

Objective

Write an appropriate linear equation for a given word problem and find the solution.

Common Core State Standards

- **7.EE.4a** Solve word problems leading to equations of the form $px + q = r$ and $p(x + q) = r$, where p, q, and r are specific rational numbers. Solve equations of these forms fluently. Compare an algebraic solution to an arithmetic solution, identifying the sequence of the operations used in each approach. *For example, the perimeter of a rectangle is 54 cm. Its length is 6 cm. What is its width?*

Expressions and Equations

Solving Linear Equations

Students have learned that a linear equation is made up of up to 2 polynomials or monomials that have variables of only the first degree. They know how to determine whether a given equation is linear and how to solve linear equations with one variable. This lesson focuses on how to solve applications of linear equations with variables on both sides of the equal sign.

Try It! *Perform the Try It! activity on the next page.*

Talk About It

Discuss the Try It! activity.

- **Ask:** *Is the amount of money the children have expressed in terms of Corina or in terms of Matsu?*
- **Ask:** *Are there any like terms you can combine on either side?*
- **Ask:** *If you take (subtract) an x-block from the left side and an x-block from the right side, will the balance hold? What if you take a second x-block from each side?*

Solve It

Reread the problem with students. **Ask:** *What should you do now that you know that the x-block is worth $3?* Help students understand that "$x = 3$" means that Matsu has $3. Since Corina has 3 more than 4 times that much, she has $15.

More Ideas

For other ways to teach about equations—

- Have students solve the problem using Algebra Tiles™.
- Extend the activity by having students solve $3x - 8 = 2 - 2x$ and $6x + 5 = 8x - 1$.

Formative Assessment

Have students try the following problem.

Solve the given equation.

$3x + 7 = 5x + 3$

A. $2x + 3 = 7$ **B.** $x = 2$ **C.** $3x + 4 = 5x$ **D.** $x = 4$

Try It! 15 minutes | Pairs

Here is a problem about writing equations.

Corina has $3 more than four times the amount of money that Matsu has. Together they have six times the amount that Matsu has. How much money does each child have?

Introduce the problem. Then have students do the activity to solve the problem. Distribute the materials.

Materials
- Algeblocks®
- BLM 7

1. Help students model the amount of money from the information given in the word problem. **Ask:** *How can you express the amount of money that Matsu has? How can you express the amount of money that Corina has? How can you express the amount of money that they have together?*

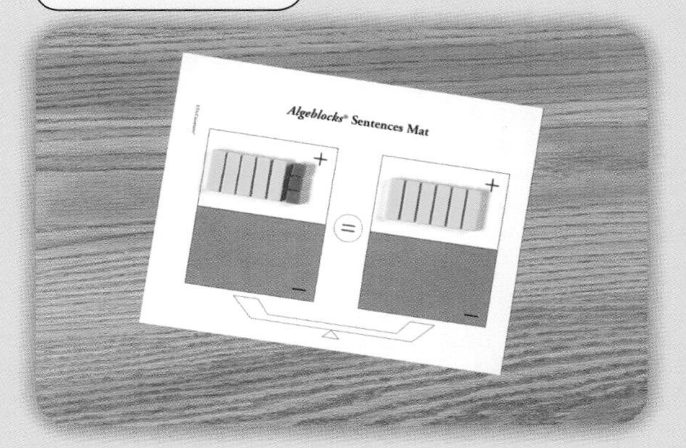

2. Have students write the equation on a piece of paper. Have students place Algeblocks on the Sentences Mat to create a model of the linear equation and show that together they have six times as much as Matsu.

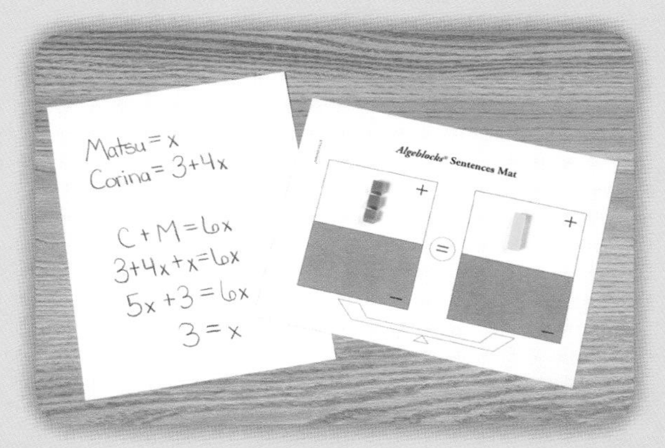

3. Ask students to take a yellow *x*-block from the left side at the same time that they take an *x*-block from the right side until their mat shows 3 green unit blocks on the left and 1 *x*-block on the right side. **Ask:** *What is the value of* x*?* Have students find the solution and write the answer on a piece of paper.

Look Out!

Some students will stop working when they have found the value of *x*. Ask them to reread the problem to see what question is being asked. **Ask:** *Does each child have $3?* Explain to students that they must take the *x*-value and apply it to the word problem in order to determine the amount of money that each child has.

Use Algeblocks and an Algeblocks Sentences Mat to model the equation shown and then solve it. Write the equation and the solution.

1. (Check students' work.)

$3x + 2 = 4x$

$x = 2$

Using Algeblocks and an Algeblocks Sentences Mat, model each equation. Sketch the model. Write each solution.

2. $3x + 9 = 4x$

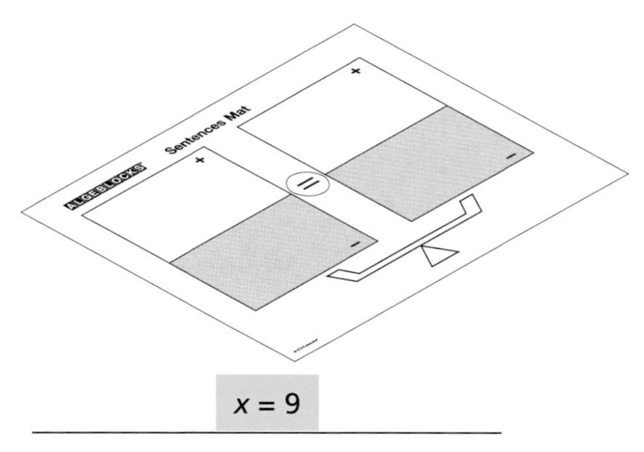

$x = 9$

3. $4x = 12 + 3x$

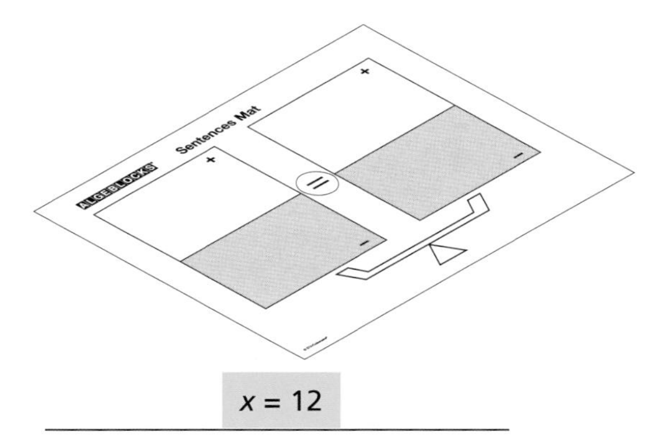

$x = 12$

Find each solution.

4. $x + 1 = 2x$

$x = 1$

5. $4x = 1 + 3x$

$x = 1$

6. $6x + 5 = 7x$

$x = 5$

7. $x + 6 = 2x$

$x = 6$

8. $10x = 9x + 9$

$x = 9$

9. $8x + 8 = 7x$

$x = -8$

Answer Key

Challenge! When solving an equation, how do you get both the variable terms on the same side of the equal sign? Explain.

Challenge: (Sample) Add the opposite of the term that is with the constant term to both sides of the equation.

Objective

Write a two-step equation to solve a real-world problem.

Common Core State Standards

- **7.EE.4a** Solve word problems leading to equations of the form $px + q = r$ and $p(x + q) = r$, where p, q, and r are specific rational numbers. Solve equations of these forms fluently. Compare an algebraic solution to an arithmetic solution, identifying the sequence of the operations used in each approach. *For example, the perimeter of a rectangle is 54 cm. Its length is 6 cm. What is its width?*

Problem Solving: Two-Step Linear Equations

Up until now, students have been working with one-step linear equations. This activity demonstrates the use of Algeblocks® as an aid to representing and solving two-step linear equations.

Try It! *Perform the Try It! activity on the next page.*

Talk About It

Discuss the Try It! activity.

- **Ask:** *How do we know that 19 = 4*b *+ 7 is the correct equation? What does the 19 represent? What does the 4*b *represent? What does the 7 represent?*
- **Ask:** *What can we remove from both sides of the equation that will still allow the equation to stay balanced?*
- **Ask:** *If 4 blocks equal 12, what does each block equal? What does the block represent?*
- **Ask:** *What does* b *represent? How many bushes did Lauren trim?*

Solve It

Reread the problem with students. **Say:** *Explain how you determined the number of bushes Lauren trimmed.* **Ask:** *How much would Lauren earn if she trimmed 6 bushes? 12 bushes?*

More Ideas

For another way to teach about two-step linear equations—

- Have students create a table of values and use the XY Coordinate Pegboard to display a graph that shows how much Lauren can earn for mowing larger lawns.

Formative Assessment

Have students try the following problem.

Solve for y.

$3y - 12 = y - 18$

A. $y = 5$ **B.** $y = -5$ **C.** $y = -3$ **D.** $y = 3$

Try It! 15 minutes | Pairs

Here is a problem about solving a two-step equation.

During the summer, Lauren mows her neighbor's lawn. She is paid $7 for mowing the lawn and an additional $4 for each bush she trims. If she was paid $19, how many bushes did she trim?

Introduce the problem. Then have students do the activity to solve the problem. Distribute the materials.

Materials

- Algeblocks®
- BLM 7

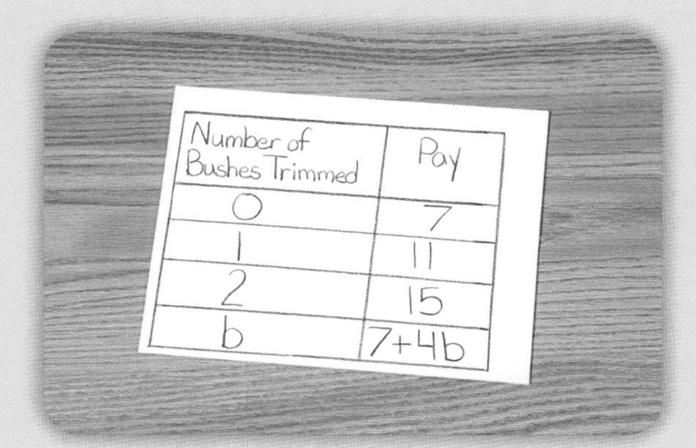

Number of Bushes Trimmed	Pay
0	7
1	11
2	15
b	7+4b

1. Have students create a table that reflects the information given in the story problem. Use this table to generate an equation.

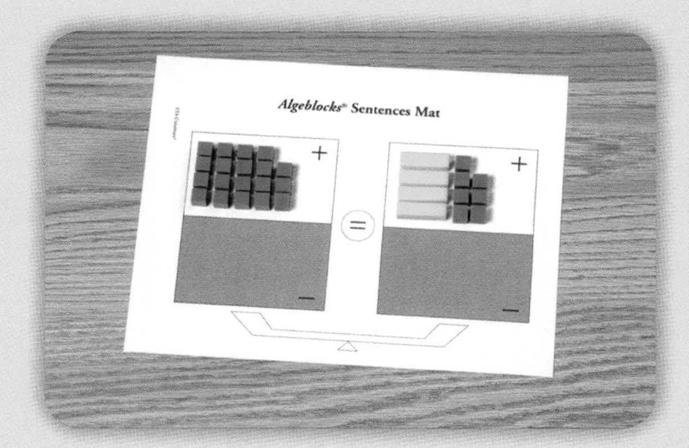

2. Have students represent this equation with Algeblocks, using the yellow *x* blocks to represent the number of bushes (*b*) Lauren trimmed and the green unit blocks for dollar amounts.

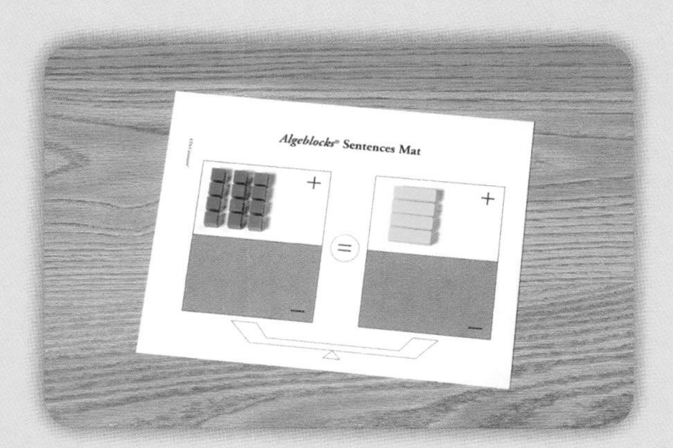

3. Have students remove 7 unit blocks from each side by making zero pairs, leaving them with 12 unit cubes on the left and 4 *b* blocks on the right. These should be arranged in equal groups as shown. From here, have students divide both sides by the number of equal groups (4). The result is $b = 3$. Remind students that *b* represents the number of bushes that Lauren trimmed.

⚠ Look Out!

Watch for students who do not reverse the order of operations when solving the equation. Remind students that when solving equations they are "undoing" the operations.

Use Algeblocks and an Algeblocks Sentences Mat to model the equation shown. Write the equation. Write the equation after the first step and write the solution.

(Check students' work.)

1.

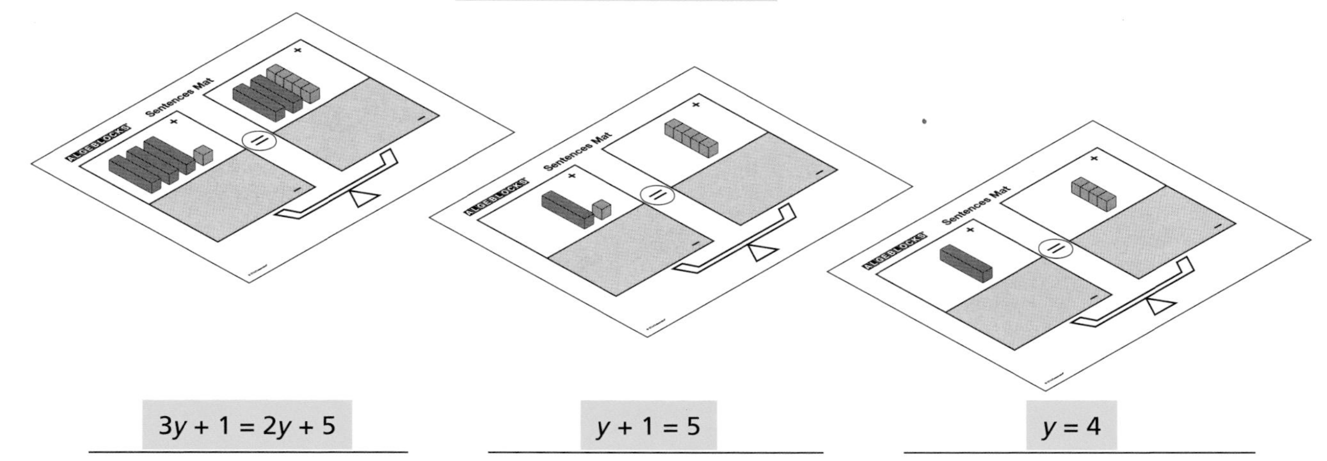

$3y + 1 = 2y + 5$ $y + 1 = 5$ $y = 4$

Using Algeblocks and an Algeblocks Sentences Mat, model the equation. Sketch the model, the first step, and the solution.

2. $2x + 9 = 13$

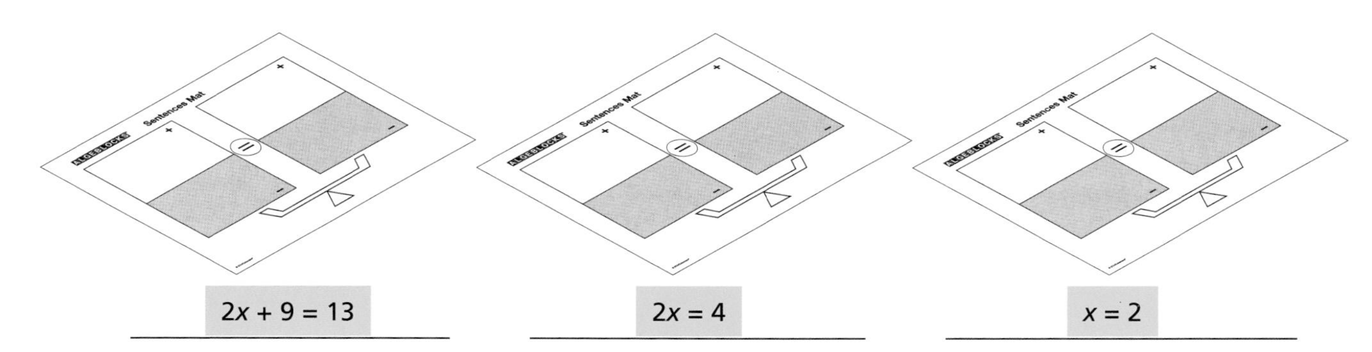

$2x + 9 = 13$ $2x = 4$ $x = 2$

Solve each equation.

3. $4x + 10 = 9x$

$x = 2$

4. $5x = 12 + x$

$x = 3$

5. $6y + 10 = 8y$

$y = 5$

6. $2x + 12 = 5x$

$x = 4$

7. $10y = 6y + 8$

$y = 2$

8. $4y + 3 = 7y$

$y = 1$

Answer Key

Challenge! Describe the two steps you used to solve the equations on the previous page. Choose an equation, show the step, and explain the reason for each step.

Challenge: (Sample) The first step is to add or subtract one of the constant terms from both sides. The second step is to divide both sides of the equation by the coefficient on the variable term.

Geometry

In seventh grade, students reason about relationships among two-dimensional figures using scale drawings and informal geometric constructions. They gain familiarity with the relationships between angles formed by intersecting lines and understand the characteristics of angles that create triangles. They think about questions such as, "What segment lengths will form a triangle?"

Additionally, students work with three-dimensional figures, relating them to two-dimensional figures by examining cross-sections. Students describe a face shape resulting from cuts made parallel and from cuts made perpendicular to the bases of right rectangular prisms and pyramids.

Students at this level also solve problems involving area and circumference of a circle. **Area** is the measure of a two-dimensional space enclosed by a shape–the region inside a shape. **Circumference** is the distance around a circle–the "perimeter" of a circle.

Students also solve problems involving area, volume, and surface area of two- and three-dimensional objects. **Surface area** is the total area that can be measured on an entire three-dimensional **surface**–for example, the sum of the areas of a polyhedron's faces. **Volume** is the space filled, or occupied, by a three-dimensional object.

The Grade 7 Common Core State Standards for Geometry specify that students should–

- Draw, construct, and describe geometrical figures and describe the relationships between them.
- Solve real-life and mathematical problems involving angle measure, area, surface area, and volume.

The following hands-on activities in geometry enable teachers to help students realize that memorizing a formula does not mean comprehending the formula. Teachers will help students understand that knowing why a formula works is more important than memorizing it. For example, at this level, students learn the formulas for determining the area and circumference of a circle and use those formulas to solve problems. With a hands-on understanding of the formulas, students can readily recall or even generate the formulas as needed.

Geometry

Contents

Objective

Construct figures using scale factors of 2 and 3.

Common Core State Standards

- **7.G.1** Solve problems involving scale drawings of geometric figures, including computing actual lengths and areas from a scale drawing and reproducing a scale drawing at a different scale.

Geometry

Scale Factor

The concept of proportionality is a key concept for students in the 7th and 8th grades. Students are expected to develop facility with ratios, rates, and proportions. This includes solving problems related to scale and the properties of similar figures. Students also need to develop their computational skills with rational numbers.

Try It! *Perform the Try It! activity on the next page.*

Talk About It

Discuss the Try It! activity.

- **Ask:** *Which AngLegs piece will complete the right triangle?*
- **Ask:** *What are the dimensions of the second ramp? What is the scale factor when compared to the first ramp?*
- **Ask:** *What are the dimensions of the third ramp? What is the scale factor when compared to the first ramp?*
- **Ask:** *What is the scale factor if you compare the third ramp to the second ramp that Jonathan and his friends built?*

Solve It

Reread the problem with the students. Ask students to describe the relationship between the first and second triangles. Have students write a paragraph identifying what they know about the lengths of the sides of the three triangles and what they know about the angles of the triangles.

More Ideas

For other ways to teach about scale factors—

- Students can use the XY Coordinate Pegboard to "grow" triangles with a given scale factor. They can also create square and rectangular structures on the board.
- Challenge students to build additional similar triangle triples with other AngLegs® pieces. (They do not have to be right triangles.) Have them compute the dimensions of the sides of the similar triangles.
- Have students make a large triangle on a Geoboard. Tell them to add a segment (rubber band) parallel to one side of the original triangle. Have them measure the corresponding sides of the two triangles and find the scale factor. Caution students that the scale factor will probably not be a whole number.

Formative Assessment

Have students try the following problem.

Amanda is building a $\frac{1}{12}$ scale model of an ultralight airplane. If the actual airplane has a wingspan of 30 feet, what will the wingspan of the model be?

A. 2.5 feet **B.** 3 feet **C.** 4 feet **D.** 12 feet

Try It! 30 minutes | Pairs

Here is a problem about scale.

Jonathan and his friends are designing and building a dirt bike course. They want to construct three takeoff ramps that are different sizes but that are similar in shape. The length (slanted portion) of the first, and smallest, ramp will be 1.87 meters. What will be the lengths of the second and third ramps if the friends use scale factors of 2 and 3 to build them?

Introduce the problem. Then have students do the activity to solve the problem. Distribute the materials.

Materials

- AngLegs® (at least 3 green, 3 yellow, and 6 orange)

1. Say: *Construct a right angle with a green AngLegs piece as the base and an orange piece as the "upright."* Have students find the AngLegs piece that completes the right triangle (yellow). **Say:** *Now you have constructed the smallest ramp.*

2. Say: *Extend each of the sides of the ramp with AngLegs pieces of the appropriate color.* Be sure students realize that they will need two orange pieces to form the "upright" of the larger triangle. **Say:** *The scale factor of this larger triangle is 2 because each side of the triangle is two times the original.* Remind students that the dirt bike ramps are like these similar triangles.

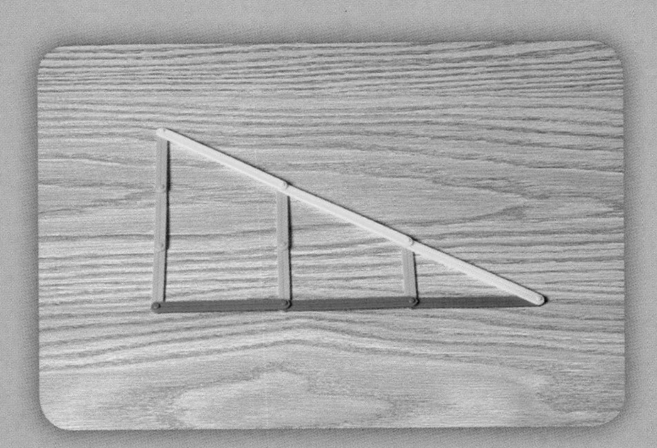

3. Say: *Extend each side again by adding another AngLegs piece of the same color as the others on that side.* This forms another, larger triangle whose dimensions have a scale factor of 3.

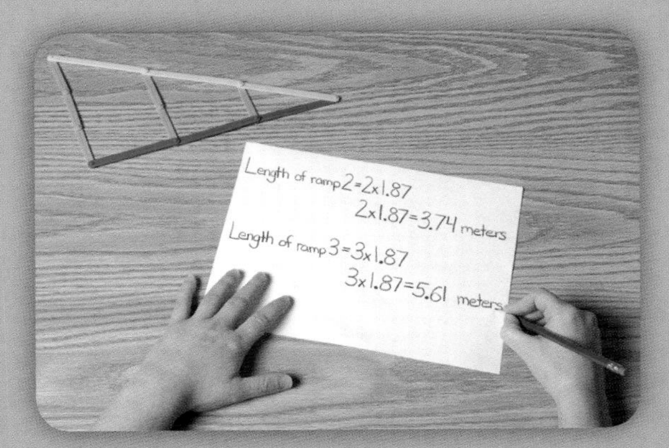

4. Say: *You have constructed models of three different, but similar, ramps.* **Ask:** *In the story problem, what is the length of the first ramp?* **Ask:** *What is the scale factor for the second ramp? What is its length?* **Ask:** *What is the scale factor for the third ramp? What is its length?*

Use AngLegs to model the triangles shown. Write the scale factor for Triangle 2.

(Check students' work.)

1. Original Triangle

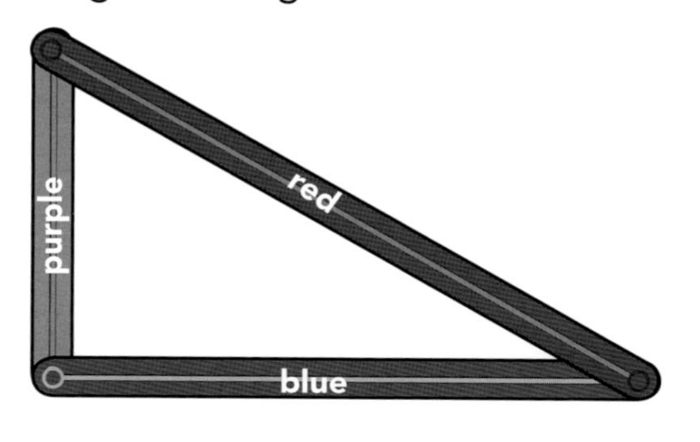

Triangle 2

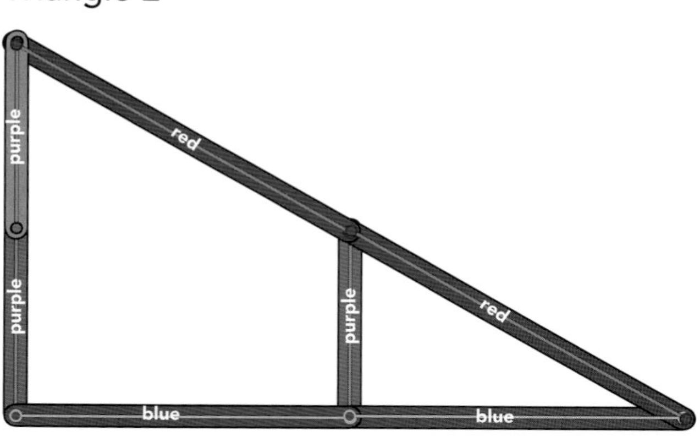

The scale factor of Triangle 2 is 2.

Using AngLegs, build a triangle with the legs named. Then build a triangle with a scale of 3:1. Sketch the models.

2. orange, yellow, and purple

Check students' models.

Draw each figure using the scale factor given.

3. scale factor of 2

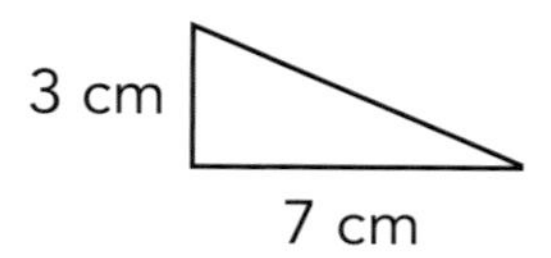

The triangle has height 6 cm and base 14 cm.

4. scale factor of 3

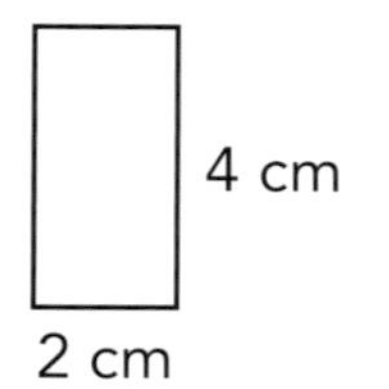

The rectangle has length 6 cm and width 12 cm.

Answer Key

Challenge! Triangle B has a scale factor of 2:1 to Triangle A. Which triangle is larger and by how much? Draw a picture.

Challenge: (Sample) Triangle B is 2 times bigger than Triangle A.

Objective

Investigate conditions for building triangles.

Common Core State Standards

- **7.G.2** Draw (freehand, with ruler and protractor, and with technology) geometric shapes with given conditions. Focus on constructing triangles from three measures of angles or sides, noticing when the conditions determine a unique triangle, more than one triangle, or no triangle.

Geometry

Construct Triangles

Triangles are polygons with three sides, and they are classified by their sides and angles. The sum of the angles of any triangle is 180°, and the sum of the lengths of any two sides in a triangle must be greater than the third side. When a figure does not meet all of these conditions, it is not a triangle. With an appropriate manipulative, students can effectively investigate conditions for building triangles. Students can determine whether a set of conditions defines one unique triangle, more than one triangle, or no triangle.

Try It! *Perform the Try It! activity on the next page.*

Talk About It

Discuss the Try It! activity.

- **Ask:** *Why was the green AngLegs® piece too short?* Elicit that the red piece was too long and/or the 45° angle was too large.
- **Say:** *Describe what happened with the yellow piece.* If necessary, explain that the yellow piece was just long enough to make a triangle.
- **Say:** *Describe what happened with the blue piece.* Elicit that the blue piece was long enough to swing through two points on the third side, so it was possible to form two triangles.
- Discuss the ways to define triangles. For example, explain that a triangle can be defined by its 3 sides, by 2 sides and the angle between them, by 1 side and it's 2 adjacent angles, and so on.

Solve It

Reread the problem with students. Have students draw the triangles they made. Have them include the angle measures—(45°, 45°, 90°), (10°, 45°, 125°), (45°, 55°, 80°). Have students answer the question in the problem. Discuss.

More Ideas

For another way to teach about constructing triangles—

- Have students use a protractor and ruler to draw triangles given certain conditions. Include conditions that lead to one triangle, two triangles, and no triangle. With an emphasis on precise measurements, students will be able to make accurate determinations.

Formative Assessment

Have students try the following problem.

Miguel measures two sides of a sail: 15 feet and 8 feet. Which could be the measurement of the third side?

A. 6 feet **B.** 10 feet **C.** 23 feet **D.** 30 feet

Try It! 25 minutes | Pairs

Here is a problem about constructing triangles.

Hannah is designing a triangular pen for her miniature play horses. She has some AngLegs to investigate different triangles. She is fixing one angle at 45° and she is fixing two of the side lengths by using a red AngLegs piece for the first side and blue, green, or yellow for the second side. Help Hannah investigate the triangles she can build. Which triangle is best?

Introduce the problem. Then have students do the activity to solve the problem. Distribute the materials.

Materials

- AngLegs® (3 of each color)
- BLM 8
- colored pencils
- straightedge

1. Have students study the BLM. Have them note the fixed angle of 45°. Have them identify where to place the red AngLegs piece and where they will be attaching the green, yellow, and blue pieces in subsequent steps. Note that the long gray side represents an unknown third side of the triangle that students will try to make.

2. Say: *Put a red AngLegs piece in its place on the diagram. Now let's choose the second side. Attach a green piece to the right end of the red piece.* Have students tell whether they are able to form a triangle. Elicit that they cannot, because the green piece is too short. **Ask:** *How would you change the 45° angle or the red side to make a triangle?*

3. Have students try a yellow piece in place of the green piece. **Ask:** *Can you make a triangle*? *Is that the only triangle you can make*? Elicit that it is. Suggest that an AngLegs piece will fit as the third side, and have students find that it is a yellow piece. Have them build the triangle and measure and record all the angles.

4. Have students try a blue piece in place of the yellow piece. **Ask:** *Can you make a triangle*? *How many*? Elicit that two different triangles can be made. **Ask:** *What would you do to the 45° angle so that only one triangle could be made? So that no triangle could be made?*

Use the AngLegs shown. Determine whether you can build a triangle.

1. (Check students' work.)

Can you build a triangle? No

Using AngLegs, try to make at least one triangle. Draw the triangle(s) or write an explanation if no triangle can be made.

2. Angles: 30°, 60°, 90°

Possible; red-blue-purple and yellow-green-orange

3. Sides: orange, orange, yellow

Not possible; sum of short sides equals long side

4. Angles: 30°, 30°, 60°

Not possible; sum of angles not 180°

5. Sides: blue, green; Angle between: 45°

Possible; blue-green-green

Answer Key

Challenge! Can you define a triangle by naming its three angles? Explain.

Challenge: (Sample) No. Naming the three angles does not tell us which triangle it is. For any three angles whose sum is 180°, there are an infinite number of triangles, all a different size.

Objective

Find the circumference of a circle.

Common Core State Standards

- **7.G.4** Know the formulas for the area and circumference of a circle and use them to solve problems; give an informal derivation of the relationship between the circumference and area of a circle.

Geometry

Circumference of a Circle and π

Students look at the ratio of circumference to diameter for various circles and develop both an approximation of the value of π and the formula for finding circumference. While a single circle shows the ratio, a larger number of examples helps students recognize the consistency of the ratio and provides a stronger basis for making a generalization.

Try It! *Perform the Try It! activity on the next page.*

Talk About It

Discuss the Try It! activity.

- **Ask:** *When might it be more useful to use $\frac{22}{7}$ as an approximation for π? When might it be more useful to use 3.14 for π?*
- **Ask:** *How would you find the circumference of a circle if you know its radius?* Explain that since the diameter is twice the length of the radius, the value $2r$ can be substituted for d in the formula for finding circumference: $C = \pi d = 2\pi r$.
- **Ask:** *How can you find the diameter of a circle if you know its circumference?*

Solve It

Reread the problem with students. Have students explain how to find the circumference of a circle when the diameter is known. Then have them find the length of ribbon Kaden needs to fit exactly around the top edge of the can.

More Ideas

For other ways to teach about circumference and π—

- Have students trace the inner and outer circles of the Rainbow Fraction Circle Rings and then measure the circumferences of the traced circles to develop the concept of π.
- Provide each group with a different circular object, such as a Two-Color Counter, spinner, Deluxe Rainbow Fraction® Circles, or Relational GeoSolids® cylinder. Have each group find the ratio of circumference to diameter of their object. Record results on the board and have students generalize the ratio—that is, determine π.

Formative Assessment

Have students try the following problem.

The diameter of a circle is 52 inches. Which expression can you evaluate to find the circumference?

A. $52 \div \pi$ **B.** $\pi \div 52$ **C.** $52 \times \pi$ **D.** $26 \times \pi$

Try It! 30 minutes | Groups of 4

Here is a problem about finding the circumference of a circle.

Kaden is decorating a can for his mother to store her small crafts. He wants to glue a piece of ribbon to the top edge of the can so that it goes around the can exactly one time. How much ribbon does he need if the diameter of the can is 14 cm?

Introduce the problem. Then have students do the activity to solve the problem. Distribute the materials. Have students start a recording chart with these headings: *Object, Diameter (d), Circumference (C),* and $\frac{C}{d}$.

Materials

- Relational GeoSolids® large and small cylinder (1 set per group)
- other circular objects (optional)
- BLM 9
- paper (1 sheet per group)
- string (2 feet length per group)
- centimeter rulers (1 per group)
- calculators (1 per group)

1. Have students measure the diameter and circumference of the base of the large and small cylinders and record each measurement to the nearest tenth of a centimeter. Then have students divide to complete the chart.

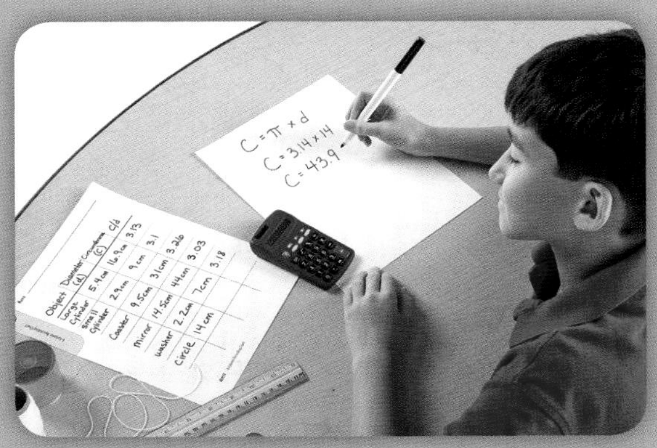

3. Ask: *How can you find the circumference of a circle if you know its diameter? What formula can you use?* Write $C = \pi \times d$ on the board. **Say:** *Add a circle with a diameter of 14 cm to your recording sheet. Use the formula to find the circumference.*

2. Have students measure the diameter and circumference of other circular objects to the nearest tenth of a centimeter and complete the table. **Ask:** *What pattern do you see in the measurements?* Write the symbol π on the board. **Say:** *This symbol is called* pi. *We often use 3.14 or* $\frac{22}{7}$ *to approximate its value.*

⚠ Look Out!

Be sure that students measure each diameter and circumference correctly. Remind them to measure the diameter at the widest part of the circle. This will help students calculate a more accurate number for π. Explain to students that π is the same for any circle, no matter how big or small. Students' calculations for π may differ slightly.

Use Relational GeoSolids to model each cylinder. Use a ruler to find the diameter of the base. Find the circumference of the base. Use 3.14 for π.

(Check students' work.)

1. 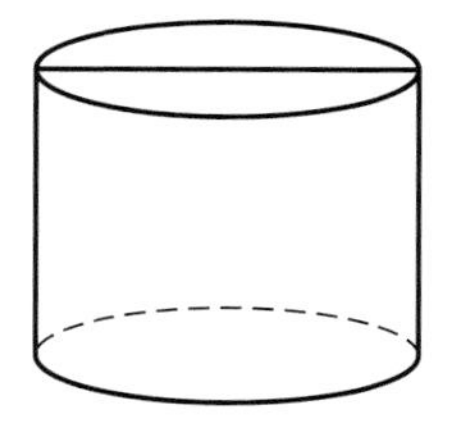

Answers will vary depending on size of Relational GeoSolids available.

2. 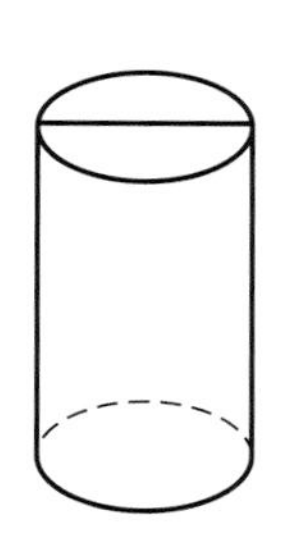

Answers will vary depending on size of Relational GeoSolids available.

Draw a circle that has each diameter. Find the circumference of the circle. Use 3.14 for π.

3. 3 inches

9.42 in.

4. 11 centimeters

34.54 cm

Find the circumference of each circle. Use 3.14 for π.

5. 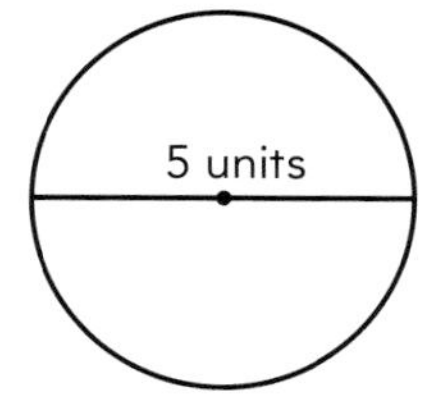

15.7 units

6. 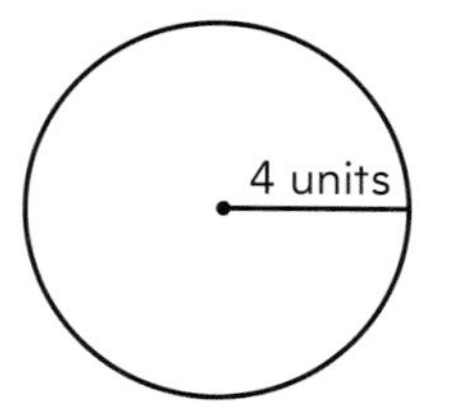

25.12 units

7. 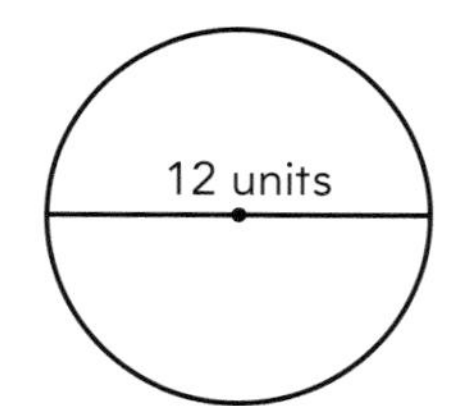

37.72 units

Answer Key

Challenge! Explain the meaning of π in terms of the parts of a circle. How is the circumference of a circle related to π?

Challenge: (Sample) The value of π is the ratio of any circle's circumference to the diameter of the circle.

Objective

Find the area of a circle.

Common Core State Standards

- **7.G.4** Know the formulas for the area and circumference of a circle and use them to solve problems; give an informal derivation of the relationship between the circumference and area of a circle.

Geometry

Area of a Circle

Measurement concepts are closely related to other mathematics topics, such as geometry and algebra. To develop and conceptualize the formula for the area of a circle, students first estimate the area by tracing a circle on grid paper. Moving the parts of the circle to form a shape that resembles the more-familiar parallelogram helps students justify and internalize the formula.

Try It! *Perform the Try It! activity on the next page.*

Talk About It

Discuss the Try It! activity.

- **Ask:** *Why is the area of the circle written in square centimeters?*
- **Ask:** *What is the relationship between the radius, diameter, circumference, and area of a circle?*
- **Ask:** *How would you find the area of a circle if you know its diameter?*

Solve It

Reread the problem with students. Have students list the information that is needed to find the area of a circle. Then have them explain how to find the area of Maya's rug.

More Ideas

For other ways to teach about the area of a circle—

- Have students use Centimeter Cubes to estimate the area of a circle. Then, using the cubes, have students estimate the radius and diameter of the circle to calculate the area. Tell students to compare the two methods for finding the area of the circle.
- Have students find the area of the circular base of a solid from a set of Relational GeoSolids®. Have students calculate the area two ways, using 3.14 and $\frac{22}{7}$ for π.

Formative Assessment

Have students try the following problem.

The radius of a circle is 10 mi. What is the area of the circle to the nearest whole number?

A. 63 mi

B. 100 sq mi

C. 314 mi

D. 314 sq mi

Try It! 20 minutes | Groups of 4

Here is a problem about the area of a circle.

Maya has a circular rug in her bedroom. What is the area of the rug if the radius is 4.4 feet?

Introduce the problem. Then have students do the activity to solve the problem. Distribute the materials. Review the terms *radius* and *diameter.* Write the symbol π on the board. Have students give the approximate value of π as a fraction and as a decimal.

Materials

- Deluxe Rainbow Fraction® Circles (1 set per group)
- BLM 1
- paper (1 sheet per group)
- calculators (1 per group)

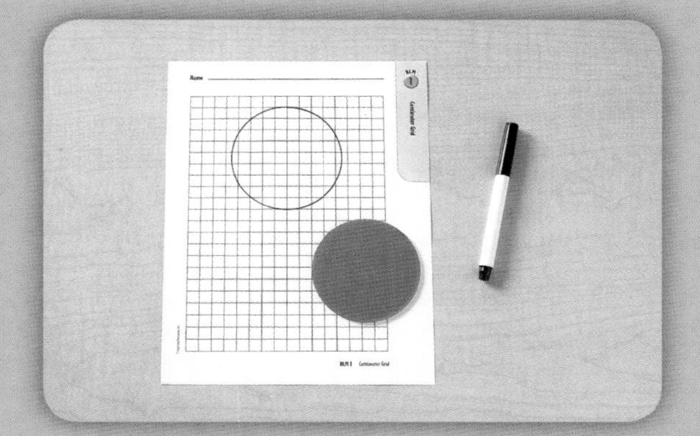

1. Have students trace the red circle on the grid paper. **Say:** *Estimate the area of the circle by counting the squares and parts of squares.* Have students share their estimates.

2. Guide students to arrange the 12 twelfths in a side-by-side pattern on the grid paper. **Ask:** *What shape does your arrangement resemble?* Write $A = b \times h$ on the board. Have students explain how to find the area of a parallelogram and have them estimate the base, height, and area of the figure.

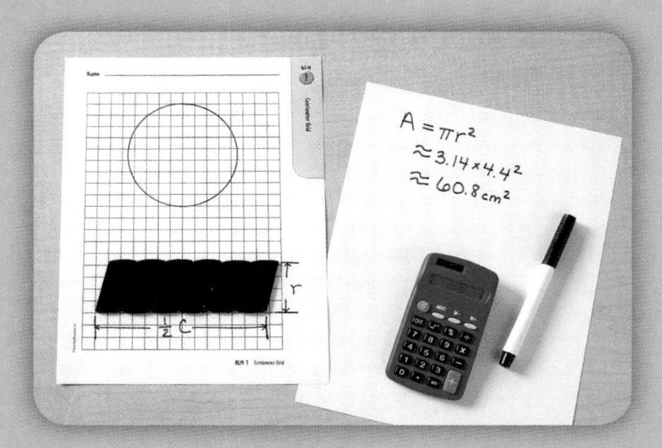

3. Ask: *What part of a circle is shown by the base of the arrangement? The height?* Show students that the base of the *parallelogram* is roughly $\frac{1}{2}$ C and that the height is roughly *r*. Write the area of the *parallelogram,* $A = \frac{1}{2} C \times r$, on the board. Replace C with $2\pi r$ and simplify to get the formula for the area of the circle, $A = \pi r^2$. Have students calculate the area of the red circle using radius 4.4 cm.

⚠ Look Out!

Some students may confuse the radius and the diameter of a circle. Have them draw and label the parts of a circle. Point out that area is always measured in square units, even when the shape has curved sides. Watch for students who think that r^2 means to multiply the length of the radius by two. Review the meaning of exponents with these students.

Use Fraction Circles to model the circle. Use a Centimeter Grid to find the area of the circle.

(Check students' work.)

1.

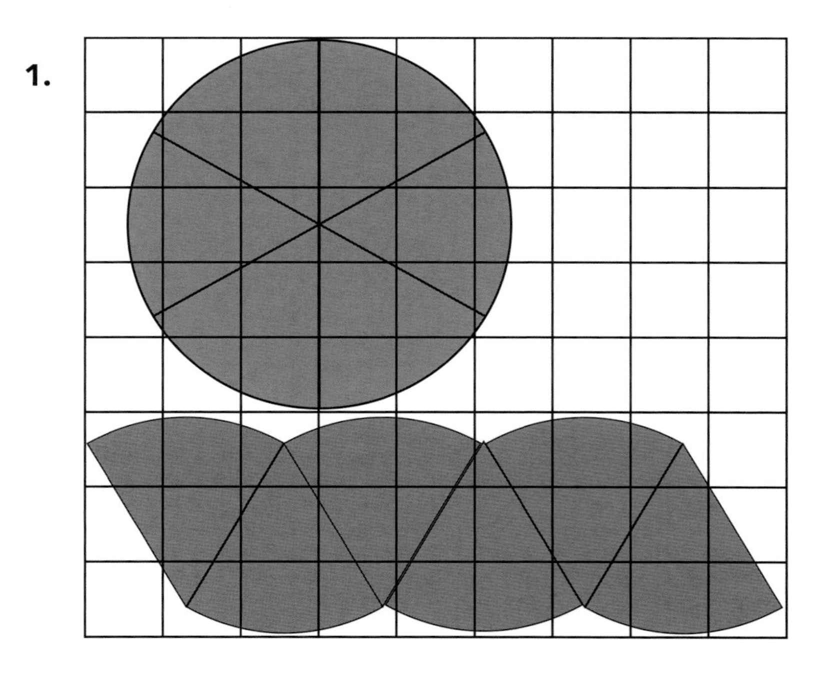

24 square units

Draw each circle described. Find the area of the circle. Use 3.14 for π.

2. 8-cm radius

200.96 square cm

3. 2-inch diameter

3.14 square in.

Find the area of each circle. Use 3.14 for π.

4.

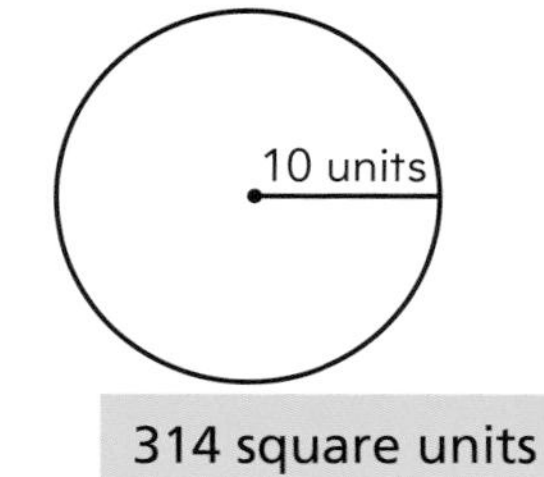

314 square units

5.

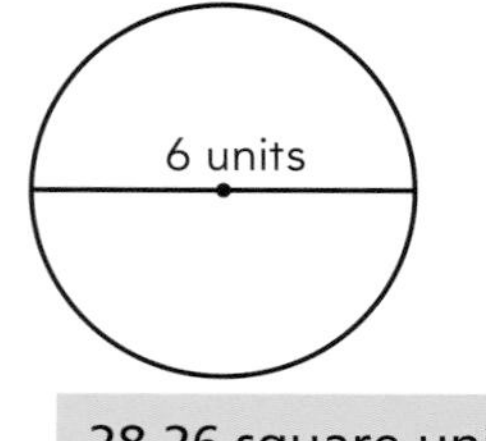

28.26 square units

6.

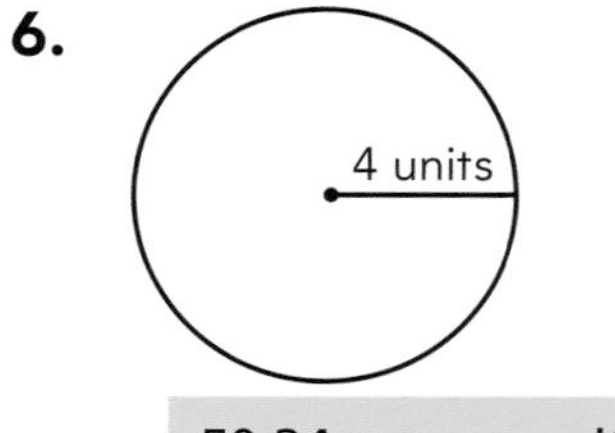

50.24 square units

Answer Key

Challenge! Determine the area of a circle on grid paper by arranging its sections into a figure having a length and width. Describe the length. Describe the width.

Challenge: (Sample) The length of the figure is one-half the circumference because half of the outer edges of the circle make up the edges of the figure. The width of the figure is equal to the radius of the circle.

Objective

Determine the area of an irregular figure by dividing it into other shapes, such as rectangles and triangles.

Common Core State Standards

- **7.G.6** Solve real-world and mathematical problems involving area, volume and surface area of two- and three-dimensional objects composed of triangles, quadrilaterals, polygons, cubes, and right prisms.

Geometry

Area of Irregular Figures

Most of the area problems students are asked to solve in 7th and 8th grade involve combining the areas of standard figures such as squares, rectangles, triangles, and some regular polygons. Finding the area of irregular figures challenges students to build on this knowledge. In addition, having students solve problems of this type can provide teachers with insight into student understanding of area and the formulas used to determine it.

Try It! *Perform the Try It! activity on the next page.*

Talk About It

Discuss the Try It! activity.

- Help students recognize the rectangle and the two triangles formed when points *ACDF* are connected by the rubber band. **Ask:** *What shapes do you see in this irregular hexagon?*
- Have students discuss how to find the area of the figure now that they see how the figure can be divided into a rectangle and two triangles. **Ask:** *What is the formula for finding the area of a rectangle? A triangle? How can we determine the area of the entire figure?*

Solve It

Reread the problem with students. If necessary, review what makes a figure regular or irregular. Have students write a short paragraph explaining how it is possible to find the area of an irregular figure by dividing it into its component shapes (rectangles and triangles).

More Ideas

For other ways to teach about the area of irregular figures—

- Have students use the XY Coordinate Pegboard to form irregular figures that have no internal rectangles and divide the figures into triangles.
- Have students make a rectangular figure with the squares from the Pattern Blocks. Instruct them to remove one square from anywhere in the figure. They should then find the area of the resulting irregular figure by thinking of it in terms of the original shape *minus* another shape.

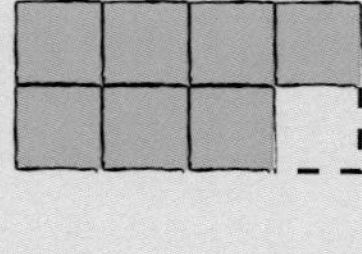

Formative Assessment

Have students try the following problem.

Find the area of the figure.

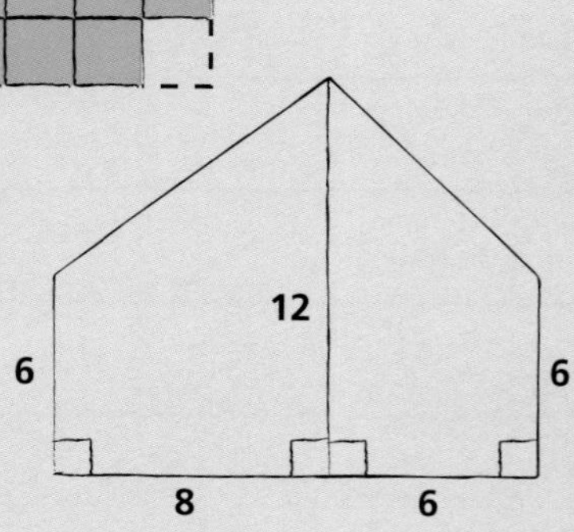

A. 84 sq. units **B.** 126 sq. units
C. 156 sq. units **D.** 168 sq. units

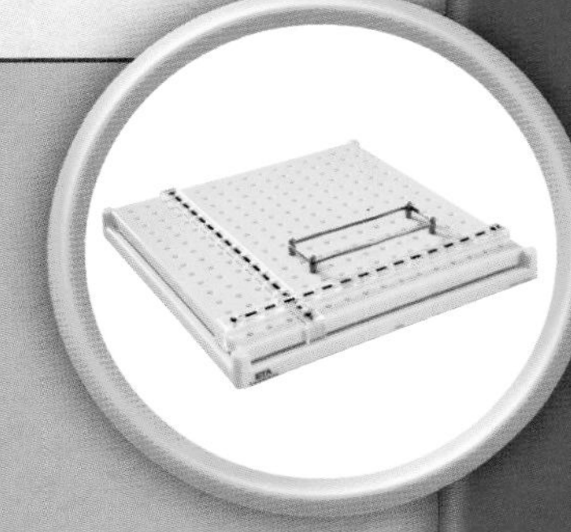

Try It! 35 minutes | Pairs

Here is a problem about the area of irregular figures.

Kristin and Erik want to install new flooring in the sunroom of their grandfather's old house. Unfortunately, the room is oddly shaped. How can they determine the area of such an irregular room?

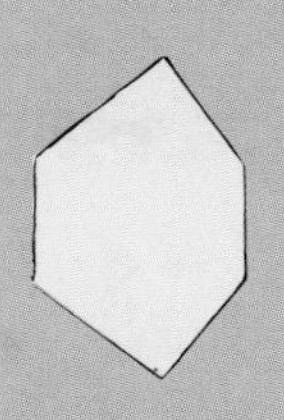

Introduce the problem. Then have students do the activity to solve the problem. Distribute the materials.

Have students set up a four-quadrant grid on their pegboard. Write the following coordinates on the board: *A* (–4, 2), *B* (–1, 7), *C* (4, 2), *D* (4, –3), *E* (2, –7), *F* (–4, –3).

Materials
- XY Coordinate Pegboard
- BLM 10
- ruler or straightedge

1. Have students place pegs at the coordinates and create figure *ABCDEF*.

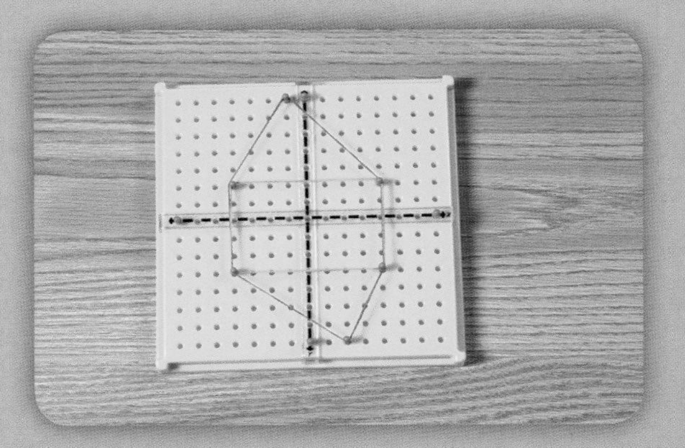

2. Have students use a rubber band to connect points *A*, *C*, *D*, and *F* to form a rectangle.

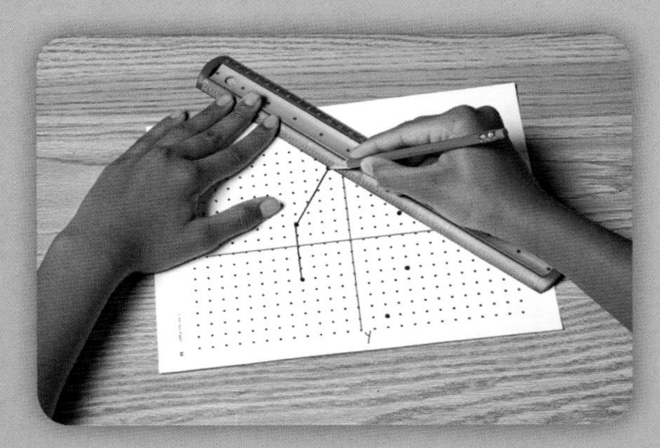

3. Have students transfer the figure to the dot paper by marking and labeling the points. Students should then connect the points with straight lines to form an irregular hexagon.

4. Have students find the area of each of the internal figures—a rectangle and two triangles. Students should then add the areas together to find the total area of the hexagon.

⚠ Look Out!

Be sure that students correctly identify the base and height of the triangles that are found inside the irregular figure. Students should realize that being able to do this is the key to solving problems of this type.

Use an XY Coordinate Pegboard to model the irregular figure. Divide the shape into triangles and a rectangle. Find the area of the irregular figure.

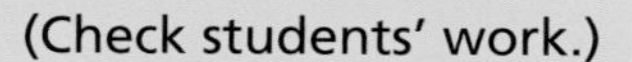

1.

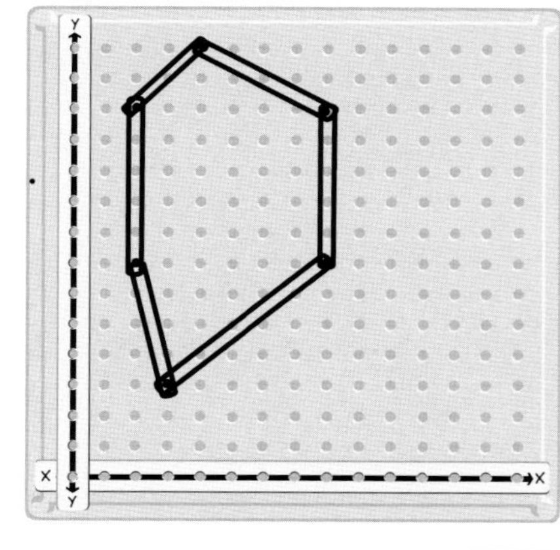

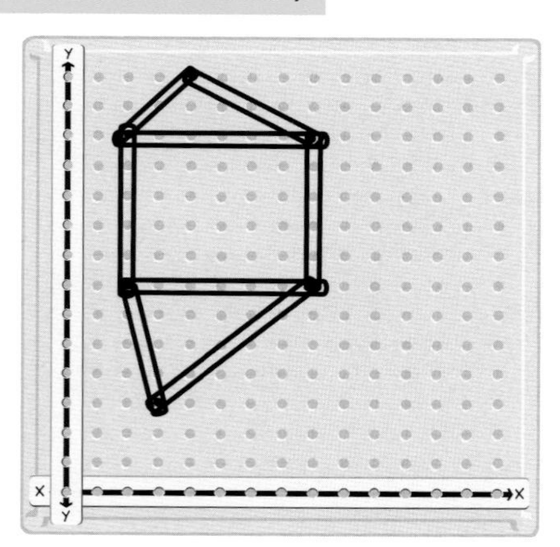

Area

triangle 6 sq units

rectangle 30 sq units

triangle 12 sq units

Area of figure 48 sq units

Using an XY Coordinate Pegboard, model an irregular figure. Sketch the model. Find the area of the irregular figure.

2.

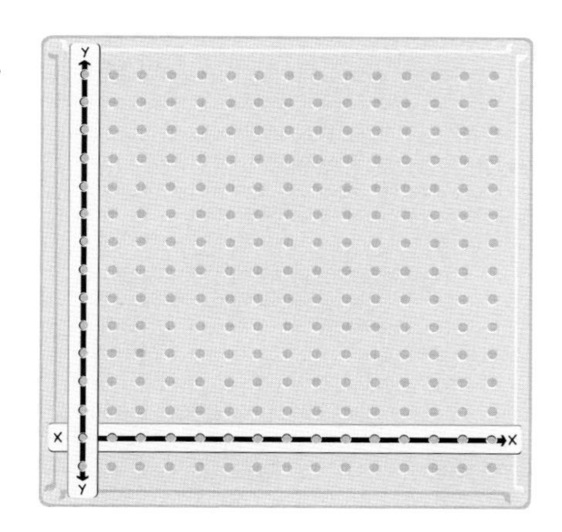

Find the areas of the shapes into which you can divide your figure.

Check students' models; answers will vary depending on models.

Area of figure ________ sq units

Find the area of each figure.

3.

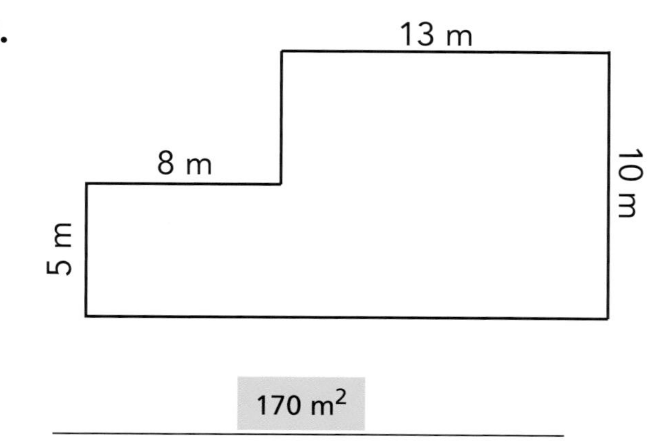

170 m^2

4.

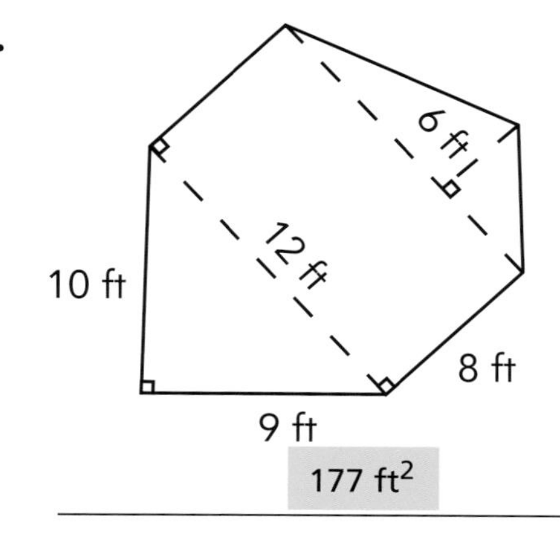

177 ft^2

Answer Key

Challenge! Why do you divide an irregular figure into other shapes to find its area? Draw a picture to help.

Challenge: (Sample) By dividing an irregular figure into common shapes, you can use formulas you know to find the area. Find the area of each shape and then add the areas together to find the area of the irregular figure.

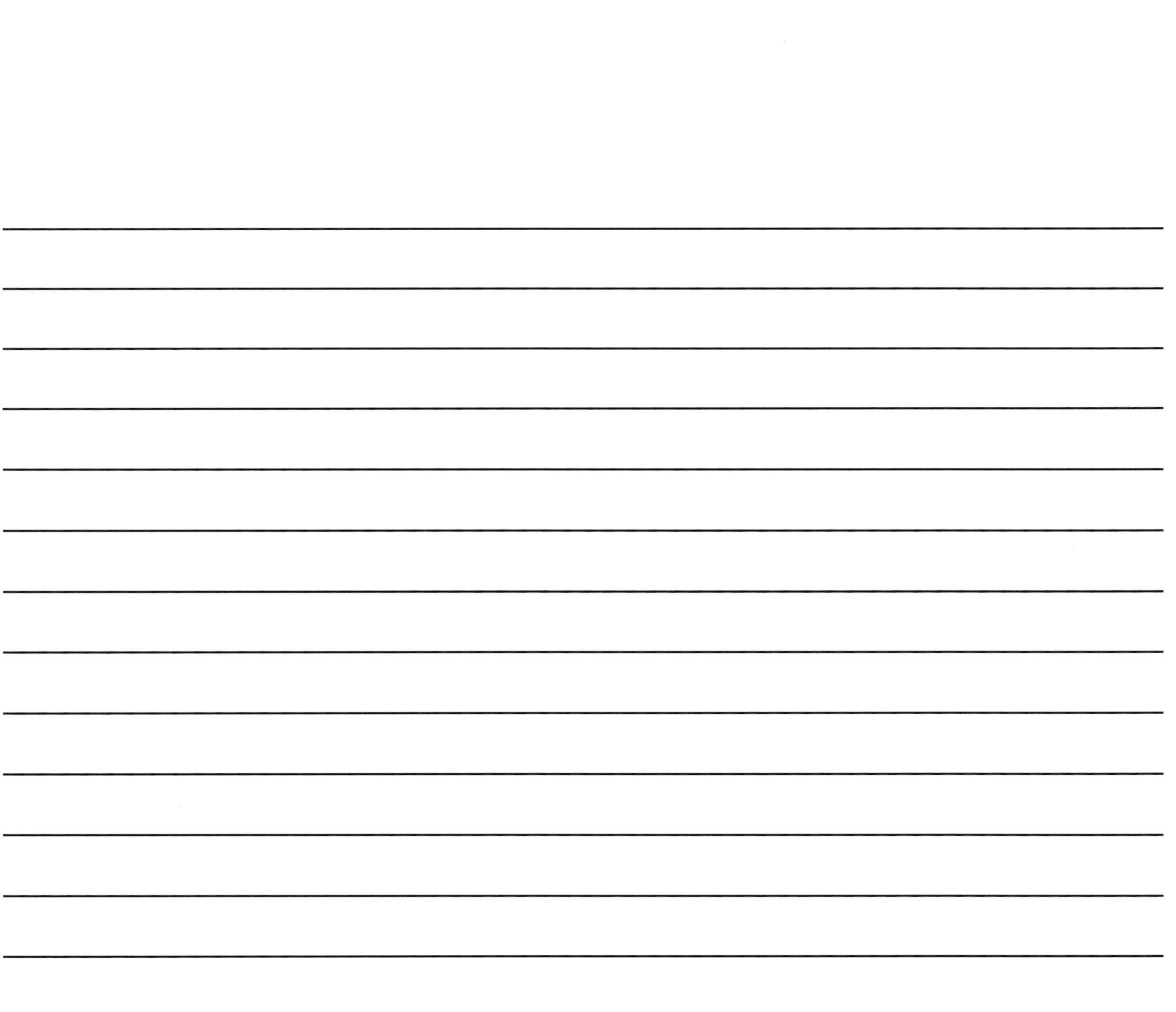

Objective

Use various polygons to form a new, larger polygon and measure its area.

Common Core State Standards

- **7.G.6** Solve real-world and mathematical problems involving area, volume and surface area of two- and three-dimensional objects composed of triangles, quadrilaterals, polygons, cubes, and right prisms.

Geometry

Polygons: Exploring Area

In the earlier grades, students used concrete objects to measure the area of several different polygons. Students will now learn how to use concrete objects to create polygons with the same area. This skill provides students with a foundation for understanding formulas and concepts that will be introduced later in their mathematics instruction.

Try It! *Perform the Try It! activity on the next page.*

Talk About It

Discuss the Try It! activity.

- **Ask:** *If two polygons have the same area, will they also have the same perimeter?*
- **Ask:** *Which of the Pattern Blocks can't be used on the triangular grid paper? Explain.*
- **Ask:** *Can you create another polygon with an area of 18 or 24 triangles?*

Solve It

Reread the problem with students. Have them explain how various polygons can have the same area. Encourage students to include sketches in their explanations. Whenever possible, encourage them to use formulas for finding the area of a given polygon.

More Ideas

For another way to teach about the area of polygons—

- Have students extend this activity by using Pattern Blocks to create several polygons, each with an area of at least 16 triangular units. Have students then count the number of triangles the shapes cover. Tell students to write the area at the top of their paper. Find several students who have created figures with the same area. Share the drawings with the class.

Formative Assessment

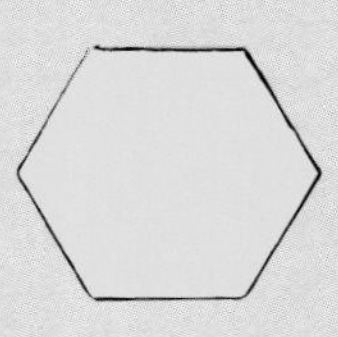

Have students try the following problem.

Which set of shapes below will not *have the same area as the figure to the right?*

A.

B.

C.

D.

Try It! 30 minutes | Pairs

Here is a problem about finding the area of specific polygons.

Samir is trying to decide which tile pattern to use in the entranceway of his home. He was told that any pattern he chooses must cover exactly 24 triangular units. He has narrowed down his choices to these three patterns:

Pattern #1: A large rhombus made up of three rows consisting of 1 triangle, 1 trapezoid, and 1 rhombus;

Pattern #2: A large parallelogram made up of three rows of 4 rhombuses;

Pattern #3: A large hexagon made up of a small central hexagon surrounded by 6 trapezoids.

Do any of the tile patterns fit the area requirements?

Introduce the problem. Then have students do the activity to solve the problem. Distribute the materials.

Materials
- Pattern Blocks
- BLM 11

1. Have students form a large rhombus consisting of three rows of 1 green triangle, 1 red trapezoid, and 1 blue rhombus as shown. **Ask:** *How many triangles are covered?*

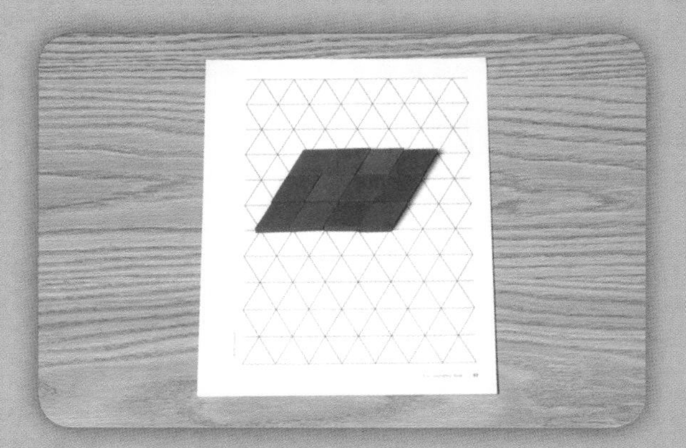

2. Now have students form a large parallelogram using three rows of 4 blue rhombuses. **Ask:** *How many triangles are covered?*

3. Have students form a large hexagon using 1 yellow hexagon and 6 red trapezoids. Have students determine how many triangles are covered. **Ask:** *Is the area the same for all three tile patterns? Do any fit Samir's requirements?*

⚠ Look Out!

Some students may think that two of the green equilateral triangles together have an area of one square inch. It is a common error since the sides of the triangles each measure 1". **Ask:** *What is the formula for the area of a triangle?* Write the formula on the board. **Ask:** *What is the base of this triangle?* Elicit "one inch." Students should notice that the height cannot also be 1" if the sloping sides are 1" long. Have students measure the base and height of the triangle if necessary.

Use Pattern Blocks and 1-inch Triangular Grid Paper to build each figure shown. Find the number of triangles covered. Write the area of the figure in triangular units.

(Check students' work.)

1.

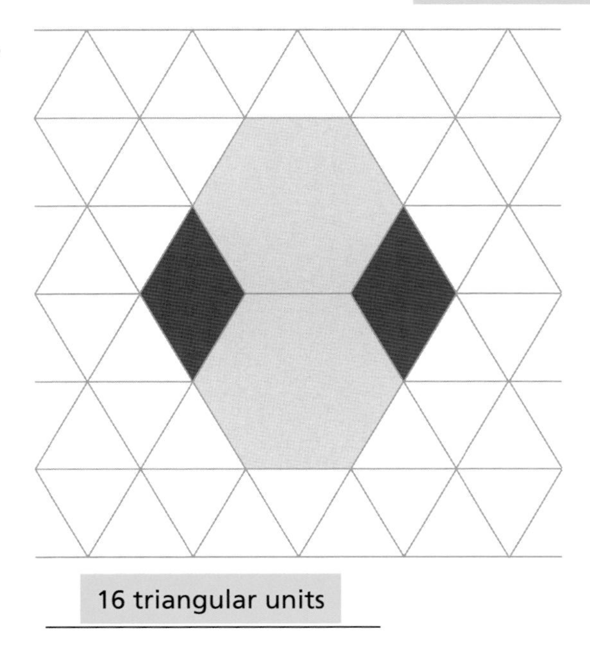

16 triangular units

2.

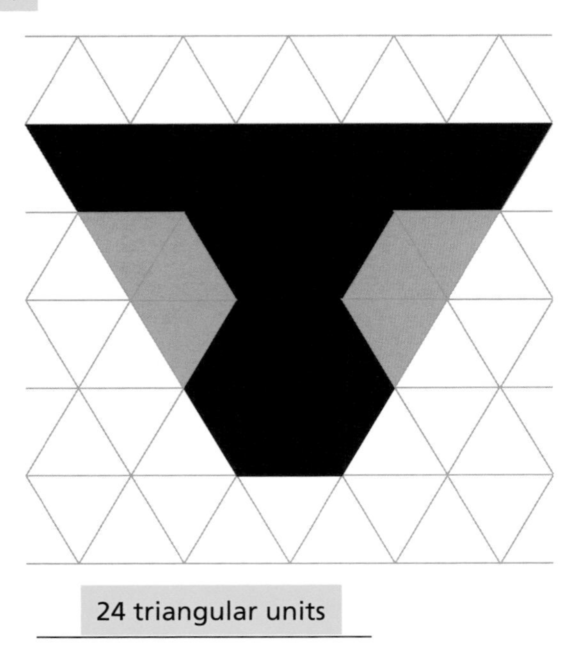

24 triangular units

Using Pattern Blocks and 1-inch Triangular Grid Paper, build a quadrilateral that has each area given. Sketch the model.

3. 20 triangular units

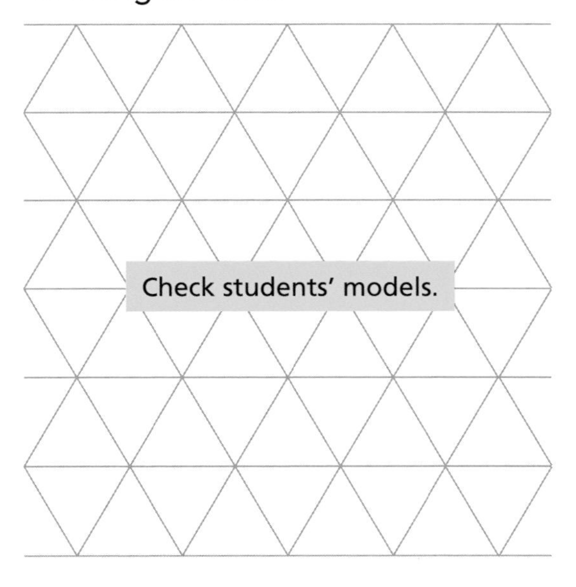

Check students' models.

4. 30 triangular units

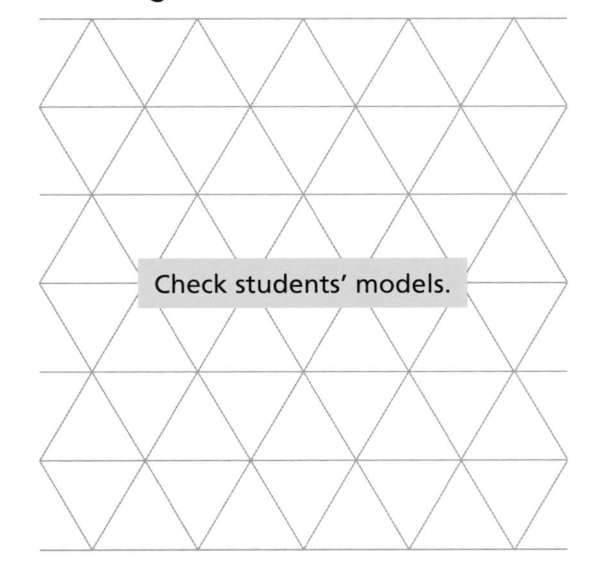

Check students' models.

Answer Key

Challenge! Explain how a hexagon formed using two trapezoids can have the same area as a hexagon formed using six equilateral triangles. Draw a picture to help.

Challenge: (Sample) The area of one trapezoid can be equal to the area of three of the equilateral triangles, so a hexagon formed using two trapezoids can have the same area as a hexagon formed by six equilateral triangles.

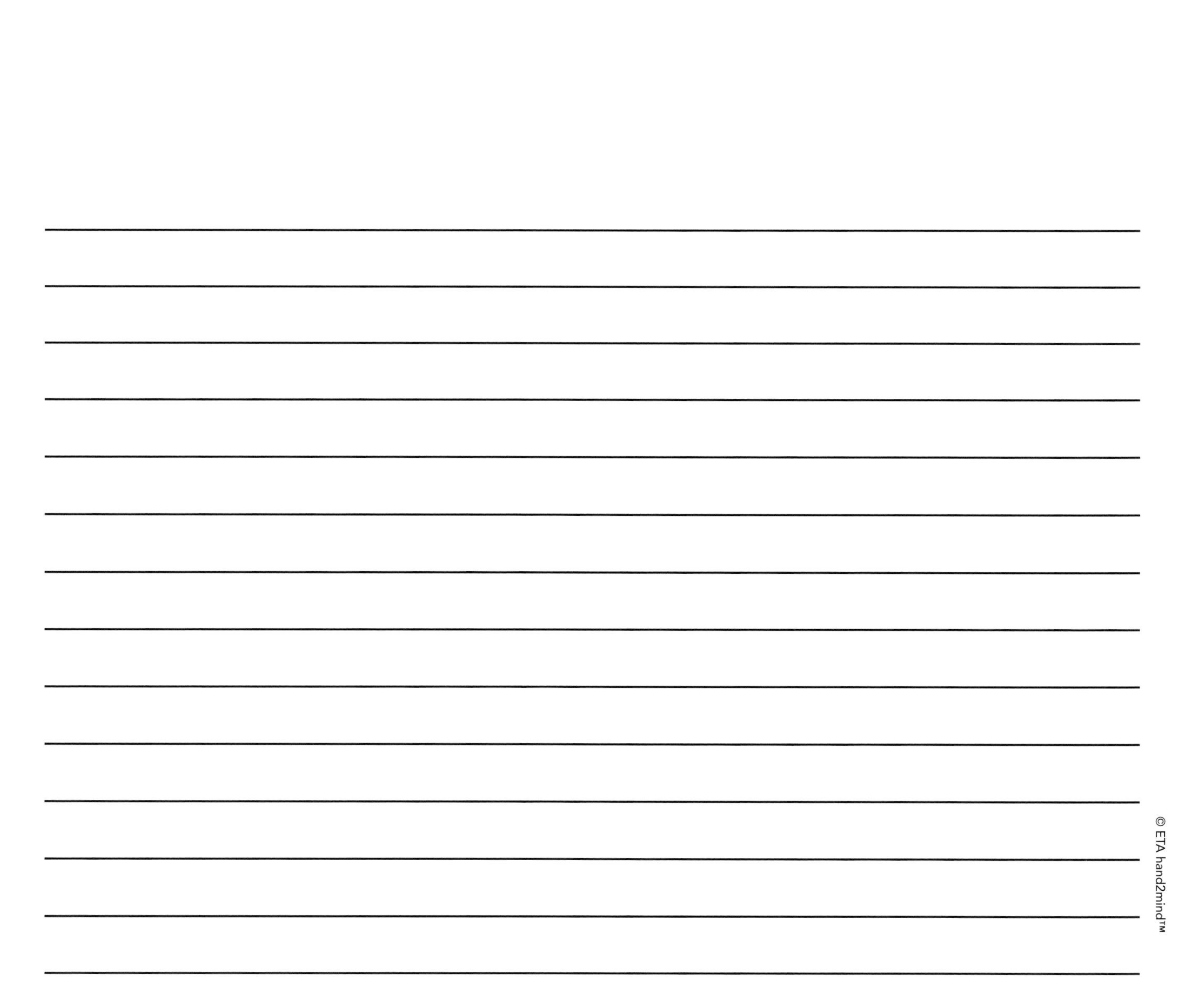

Statistics and Probability

We use statistics and probability often in our daily lives. When we hear about the "average" price per gallon of gasoline or the "chance" of snowfall, we are dealing with statistics and probability. **Statistics** is the study of how data is collected, summarized, and presented. **Probability** is the likelihood that something will happen. Together, statistics and probability are used with data to draw conclusions, and/or make predictions.

Probability is expressed mathematically as a number between 0 and 1 (e.g., a chance of rainfall of $\frac{7}{10}$, or 0.7, or 70%). A probability near 0 means an event is unlikely, a probability of $\frac{1}{2}$ means the event is neither likely nor unlikely, and a probability near 1 means the event is likely to occur. Probabilities are useful for predicting what may happen in the long term, such as realizing trends in sales, human populations, and weather.

At this level, students build on their previous work with single data distributions and address differences between populations–that is, between two data sets. Students recognize that it is difficult to gather statistics on an entire population but that a random sample can be representative of the total population and will generate valid results.

Additionally, students develop probability models and use them to find probabilities of future events (such as finding the probability that a spinner will land on a certain spot or that a penny will be tossed as a "heads" or as a "tails"). Students also find probabilities of compound events. A **compound event** consists of two or more **simple events**. Tossing a die is a simple event. Tossing two dice is a compound event.

The Grade 7 Common Core State Standards for Statistics and Probability specify that students should–

- Use random sampling to draw inferences about a population.
- Draw informal comparative inferences about two populations.
- Investigate chance processes and develop, use, and evaluate probability models.

The following hands-on activities will enable teachers to help students develop statistics and probability skills and concepts in a meaningful way. The activities will help students gain an appreciation for how mathematics can be used to model the real world.

Statistics and Probability

Contents

Objective

Estimate the size of a subgroup by sampling the larger population.

Common Core State Standards

- **7.SP.1** Understand that statistics can be used to gain information about a population by examining a sample of the population; generalizations about a population from a sample are valid only if the sample is representative of that population. Understand that random sampling tends to produce representative samples and support valid inferences.
- **7.SP.2** Use data from a random sample to draw inferences about a population with an unknown characteristic of interest. Generate multiple samples (or simulated samples) of the same size to gauge the variation in estimates or predictions. *For example, estimate the mean word length in a book by randomly sampling words from the book; predict the winner of a school election based on randomly sampled survey data. Gauge how far off the estimate or prediction might be.*

Population Sampling

The media often present data from polls and sample populations—data intended to help sway or justify a variety of actions or positions. It is important for students to understand the processes associated with population sampling and the mathematics involved in analyzing the data. In this activity, students will simulate a survey of a wildlife population by collecting and analyzing a series of representative samples.

Try It! *Perform the Try It! activity on the next page.*

Talk About It

Discuss the Try It! activity.

- **Ask:** *What does the container of colored cubes represent? What does it mean to "take a sample"? When you take a sample, what data do you need to record?*
- Have the various groups of students report their findings. Then reveal that there are 165 frogs in the pond and that 15 (9.09%) of them have mutations. Have students discuss how different groups of researchers might get different results, even though everyone followed the same directions.

Solve It

Reread the problem with students. Ask them to write to Chen explaining their procedure for answering his question. Have them include their results.

More Ideas

For other ways to teach about population sampling—

- Encourage students to look for examples of the use of population sampling (wildlife studies, election polls, and so forth) in the media. Whenever possible, discuss the sample size and number of samples taken in relation to the size of the total population.
- Have students repeat the exercise. This time, however, they should take 10 samples instead of four. Are the results the same? **Ask:** *Do you think your results are likely to be more accurate this time, less accurate, or about the same? Explain.*

Formative Assessment

Have students try the following problem.

In a sampling of 63 squirrels observed in a forest preserve, 9 were red squirrels. If the total squirrel population is estimated to be 550, approximately how many of them are red squirrels?

A. 14 **B.** 80 **C.** 470 **D.** 540

Try It! 40 minutes | Groups of 4

Here is a problem about population sampling.

Chen has noticed that some of the frogs in a nearby pond have malformed legs. He wants to know what percentage of the frog population has these mutations, but he realizes that he can't catch and examine every frog. How can Chen determine the percentage of mutant frogs in the pond without counting every frog?

Introduce the problem. Then have students do the activity to solve the problem. Distribute the materials.

Materials

- Centimeter Cubes (165 cubes, 15 of which are red)
- $\frac{1}{4}$-cup measuring cup

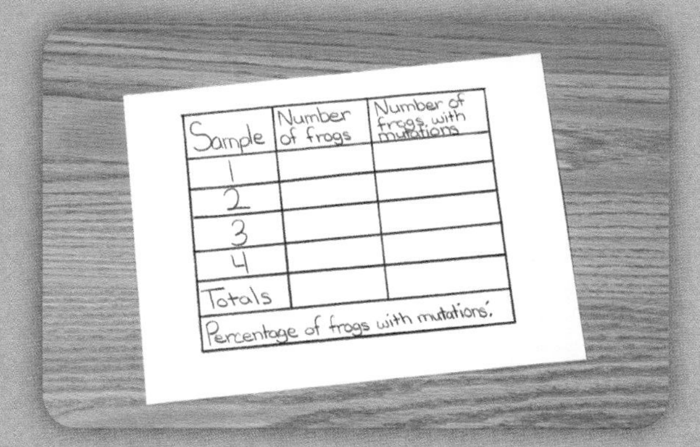

1. Explain to students that they are going to simulate a sampling process. Tell them that the red cubes will represent the mutant frogs. Have students set up a table for recording their data. The table should accommodate four samples.

2. Have students scoop out a sample of Centimeter Cubes with the measuring cup, count them, and record the total number of cubes and the number of red cubes in the sample. Then have them return the cubes to the container and mix them together. (You may explain that this is the equivalent of "catch and release.") Have students repeat the process three more times.

3. Have students calculate the total number of frogs captured and the total number of mutant frogs in all four samples. **Ask:** *According to your research, what percentage of frogs in the pond has mutations?*

Look Out!

Remind students as necessary to return the sample to the container and mix the cubes thoroughly before taking another sample.

Some students will report mutation rates of more or less than 9 percent. Have them confirm that they followed the procedure and that their calculations are correct. Reassure them that a sampling is not a literal count of every individual in the population and that samples will vary.

Use Centimeter Cubes to represent votes from a subgroup of a larger population. In the sample shown, the red cubes are modeled by the dark cubes and represent a *yes* vote. Record your results.

(Check students' work.)

1. Three samples are shown. Complete the table.

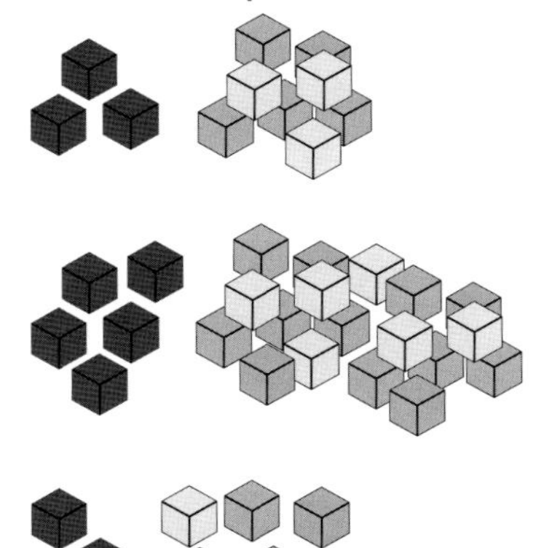

Sample	Number of Votes	Number of *Yes* Votes	Number of *No* Votes
1	11	3	8
2	23	5	18
3	9	2	7

How many votes were cast based on the samples? 43

How many votes were *yes*? 10

What percent of the votes cast were *yes* votes? 23.3%

Using Centimeter Cubes, represent votes from a subgroup of a larger population. Take three samples from a large pile of cubes. Choose a color to represent *yes* votes. Record your results.

2.

Sample	Number of Votes	Number of *Yes* Votes	Number of *No* Votes
1			
2	Answers will vary.		
3			

How many votes were cast based on the samples? ________

How many votes were *yes*? ______

What percent of the votes cast were *yes* votes? ______________

3.

Sample	Number of Votes	Number of *Yes* Votes	Number of *No* Votes
1			
2	Answers will vary.		
3			

How many votes were cast based on the samples? ________

How many votes were *yes*? ______

What percent of the votes cast were *yes* votes? ______________

Answer Key

Challenge! Why it is necessary to use a sample when seeking results from a large poplulation?

Challenge: (Sample) A population is often too large to test or survey. A sample can be tested or surveyed and the results can be used to generalize the results of the larger population.

Objective

Build a spinner that models a set of probabilities.

Common Core State Standards

- **7.SP.5** Understand that the probability of a chance event is a number between 0 and 1 that expresses the likelihood of the event occurring. Larger numbers indicate greater likelihood. A probability near 0 indicates an unlikely event, a probability around 1/2 indicates an event that is neither unlikely nor likely, and a probability near 1 indicates a likely event.

Statistics and Probability

Modeling Probability: Building Spinners

A spinner is a convenient way to model probability—the likelihood that a specified event will occur. Probability can be expressed as a fraction, decimal, or percent. In this lesson, students will build spinners to model probabilities expressed as fractions.

Try It! *Perform the Try It! activity on the next page.*

Talk About It

Discuss the Try It! activity.

- Discuss with students the fractions they were given in the problem and have students tell you what the fractions represent.
- Have students write an equation for each spinner showing the sum of the probabilities. Elicit that the sum is always 1. **Ask:** *Can a probability ever be greater than 1?* Elicit that probability is always between 0 and 1. Explain that a value near 0 means an event is unlikely, that a value near 1 means the event is likely, and that a value near $\frac{1}{2}$ means the event is neither unlikely nor likely.

Solve It

Reread the problem with students. Have them use unit fractions or a combination of unit fractions to create the spinners as described. Have students use the rings to check the accuracy of the spinners and to help them draw the spinners on a piece of paper.

More Ideas

For other ways to teach about modeling probability—

- Have students extend this activity by creating a spinner that represents the following probabilities: $P(\text{color 1}) = \frac{1}{10}$; $P(\text{color 2}) = \frac{2}{5}$; $P(\text{color 3}) = \frac{1}{5}$; $P(\text{color 4}) = \frac{3}{10}$.
- Have students design a number cube with the following probabilities: $P(1) = \frac{1}{6}$; $P(2) = \frac{1}{2}$; $P(3) = \frac{1}{3}$.

Formative Assessment

Have students try the following problem.

Which of the following probabilities is correct for the spinner shown here?

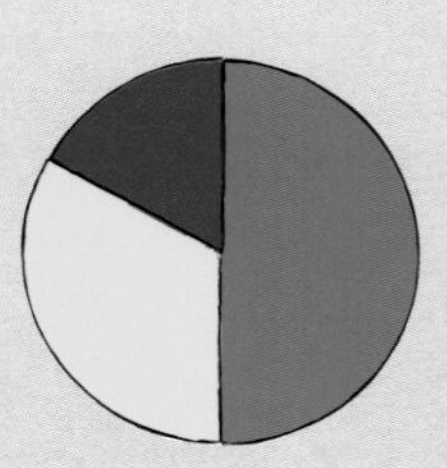

A. $P(\text{green}) = \frac{1}{3}$ **B.** $P(\text{yellow}) = \frac{1}{2}$

C. $P(\text{yellow}) = \frac{1}{6}$ **D.** $P(\text{green}) = \frac{1}{2}$

Try It! 30 minutes | Groups of 4

Here is a problem about building spinners to model probabilities.

Jake found an old board game in the attic that he'd like to play. The problem is that all of the spinners for the game are missing. Luckily, he found a description of them in the directions. He needs to replace the three spinners described here:

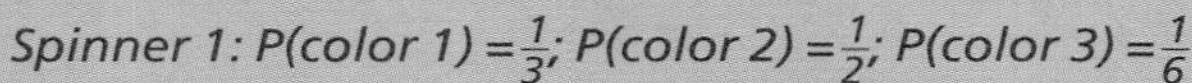

Spinner 1: P(color 1) = $\frac{1}{3}$*; P(color 2) =* $\frac{1}{2}$*; P(color 3) =* $\frac{1}{6}$

Spinner 2: P(color 1) = $\frac{5}{12}$*; P(color 2) =* $\frac{1}{4}$*; P(color 3) =* $\frac{1}{3}$

Spinner 3: P(color 1) = $\frac{3}{10}$*; P(color 2) =* $\frac{1}{5}$*; P(color 3) =* $\frac{1}{2}$

Help Jake accurately re-create the spinners he will need.

Introduce the problem. Then have students do the activity to solve the problem. Distribute the materials.

Materials

- Deluxe Rainbow Fraction® Circles
- Rainbow Fraction Circle Rings
- drawing compass
- colored markers

1. Have students build Spinner 1, using the appropriate Fraction Circles to represent each probability. **Ask:** *What color fraction piece will you use to represent* $\frac{1}{3}$*?*

2. Next have students build Spinners 2 and 3, using the appropriate Fraction Circles. **Ask:** *How will you represent* $\frac{5}{12}$*?*

3. Have students use the Measurement Ring to confirm that the spinner sections represent the correct probabilities. Have them use the ring as an aid to draw the spinners. Students may use markers to color in the sections. Students should label each section with the appropriate fraction.

⚠ Look Out!

Some students might struggle to associate the value of a probability with the idea of an event being likely, unlikely, or neither. Use Fraction Towers with a 0–1 number line to demonstrate the concept.

Use Fraction Circles and the Measurement Ring to make a spinner for the probabilities given.

(Check students' work.)

1.

Color	Probability
Black	$\frac{1}{12}$
Gray	$\frac{2}{3}$
White	$\frac{1}{4}$

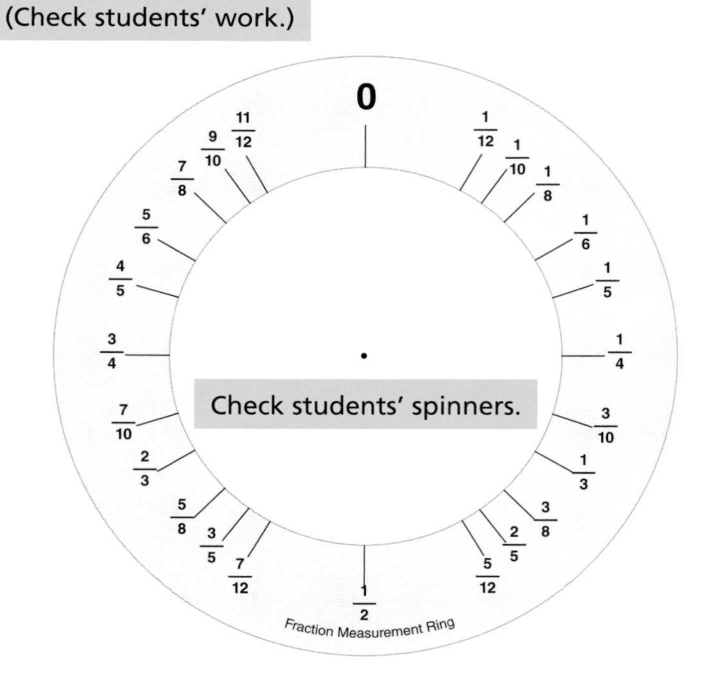

Using Fraction Circles and the Measurement Ring, make a spinner for the probabilities given.

2.

Pattern	Probability
dotted	$\frac{1}{4}$
striped	$\frac{1}{8}$
clear	$\frac{3}{8}$
solid	$\frac{1}{4}$

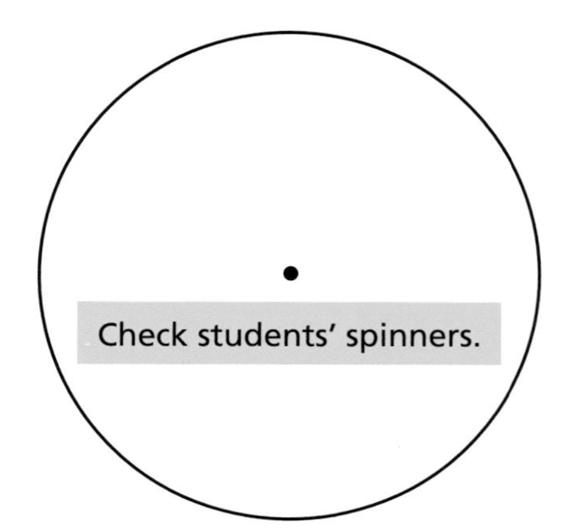

3.

Letter	Probability
A	$\frac{3}{10}$
B	$\frac{2}{5}$
C	$\frac{1}{10}$
D	$\frac{1}{5}$

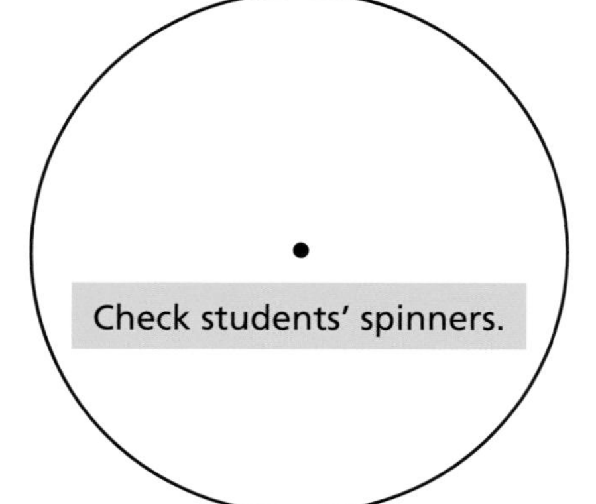

Answer Key

Challenge! When making a spinner showing certain probabilities, what must the sum of the probabilities equal? Explain. Draw a picture to help.

Challenge: (Sample) The probabilities must have a sum of 1 because together they represent the whole circle.

Objective

Find the theoretical and experimental probabilities of an event involving a spinner.

Common Core State Standards

- **7.SP.5** Understand that the probability of a chance event is a number between 0 and 1 that expresses the likelihood of the event occurring. Larger numbers indicate greater likelihood. A probability near 0 indicates an unlikely event, a probability around 1/2 indicates an event that is neither unlikely nor likely, and a probability near 1 indicates a likely event.
- **7.SP.6** Approximate the probability of a chance event by collecting data on the chance process that produces it and observing its long-run relative frequency, and predict the approximate relative frequency given the probability. *For example, when rolling a number cube 600 times, predict that a 3 or 6 would be rolled roughly 200 times, but probably not exactly 200 times.*
- **7.SP.7a** Develop a uniform probability model by assigning equal probability to all outcomes, and use the model to determine probabilities of events. *For example, if a student is selected at random from a class, find the probability that Jane will be selected and the probability that a girl will be selected.*

Theoretical and Experimental Probability with Spinners

Students learn that they can use theoretical probability to predict the results of an experiment and that it may or may not be the same as the experimental probability. Here they use spinners to learn how to distinguish and compare theoretical and experimental probability and how to express the probability as a fraction, decimal, or percent.

Try It! *Perform the Try It! activity on the next page.*

Talk About It

Discuss the Try It! activity.

- Discuss that the likelihood that an event will occur is indicated by a number from 0 to 1. Zero means the event is impossible, and 1 means the event is certain.
- **Ask:** *Given the theoretical probability of the spinner landing on a number less than 4, how many times would you expect the spinner to land on one of these numbers in 10 spins? 30 spins? 50 spins?*
- **Ask:** *As you run more trials, what do you notice about the theoretical and experimental probabilities?*
- **Ask:** *Who do you think will earn more points, Thom or Maya? Why?*

Solve It

Reread the problem with students. Ask them to find the theoretical and experimental probabilities of Thom getting a point, and to use their experiment to compare and contrast theoretical and experimental probability.

More Ideas

For other ways to teach about theoretical and experimental probabilities—

- Give polyhedral dice to pairs of students. Have one student find the experimental probability of rolling a prime number on 10, 20, and 30 rolls while the other finds and uses the theoretical probability to determine the expected results.

Formative Assessment

Have students try the following problem.

A spinner is numbered 1 through 12. Lauren spins it 15 times and it lands on a number greater than 3 ten times. What is the theoretical probability that the spinner lands on a number greater than 3?

A. $\frac{1}{5}$ **B.** $\frac{1}{4}$ **C.** $\frac{3}{4}$ **D.** $\frac{4}{5}$

Try It! 30 minutes | Pairs

Here is a problem about theoretical and experimental probabilities.

Thom and Maya are playing a game with a spinner numbered 1–8. Thom gets a point if the spinner lands on a number less than 4. Maya gets a point if the spinner lands on 4 or greater. Compare the theoretical probability that Thom will get a point with the experimental probability using 10, 30, and 50 spins.

Introduce the problem. Then have students do the activity to solve the problem. Distribute the materials. Explain the difference between *theoretical probability*, which is a calculated number, and *experimental probability*, which describes what actually occurs in an experiment.

Materials

- Spinners
- paper (1 sheet per pair)

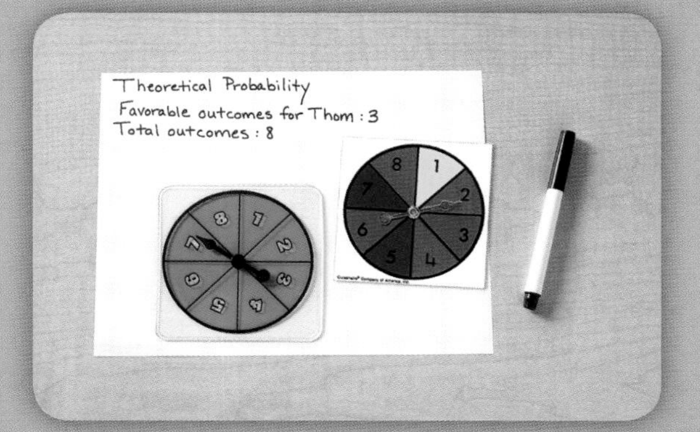

1. Say: *The theoretical probability of an event is the ratio of the number of favorable outcomes to the total number of possible outcomes.* **Ask:** *How many favorable outcomes are there for Thom? How many possible outcomes are there?*

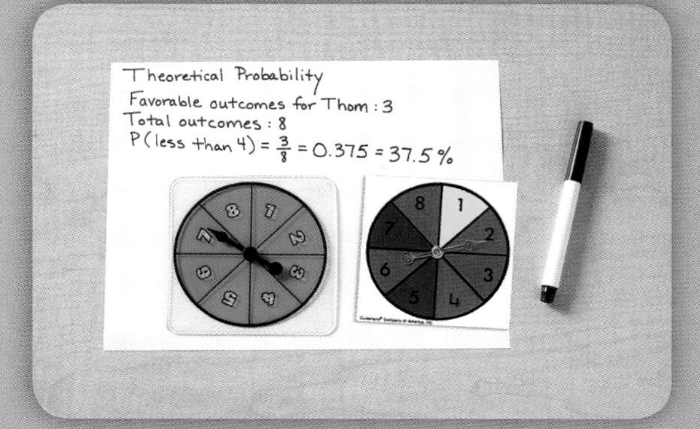

2. Say: *You can express probability as a fraction, decimal, or percent.* Help students express as a fraction the theoretical probability of the spinner landing on a number less than 4. Then have them convert the fraction to a decimal and a percent.

3. Say: *Experimental probability is the ratio of favorable trials to the total number of trials.* Have students tally favorable trials (numbers less than 4) for 10, 30, and 50 spins, and express the experimental probabilities in three ways. Then have them compare the theoretical and experimental probabilities.

⚠ Look Out!

Some students may confuse theoretical and experimental probabilities. Emphasize that theoretical probability tells what would happen if each possible outcome appears the same number of times. For example, if a spinner has five equal sections and you spin it five times, each number would appear once, or 1 out of 5 times. In an actual experiment, a number might appear more than or less than once in five spins. What actually happens is experimental probability.

Use a spinner to model probability. Find each probability. (Check students' work.)

1. $P(B)$ $\frac{1}{3}$

   $P(A)$ $\frac{1}{3}$

   $P(C)$ $\frac{1}{3}$

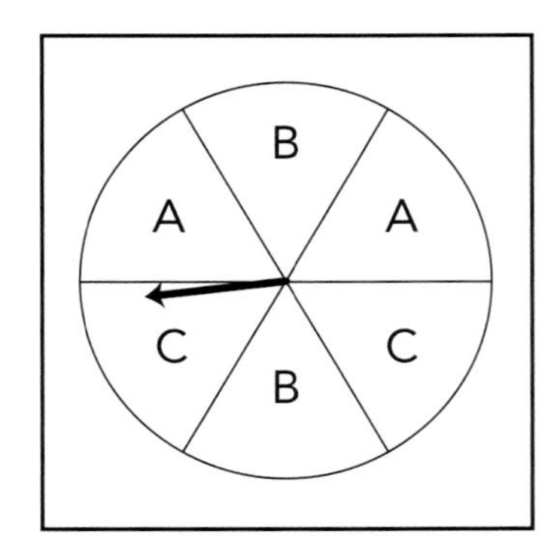

Using a spinner, model each probability. Find each probability.

2. $P(1)$ $\frac{1}{8}$

   $P(4)$ $\frac{1}{4}$

   $P(\text{number} < 5)$ 1

   $P(0)$ 0

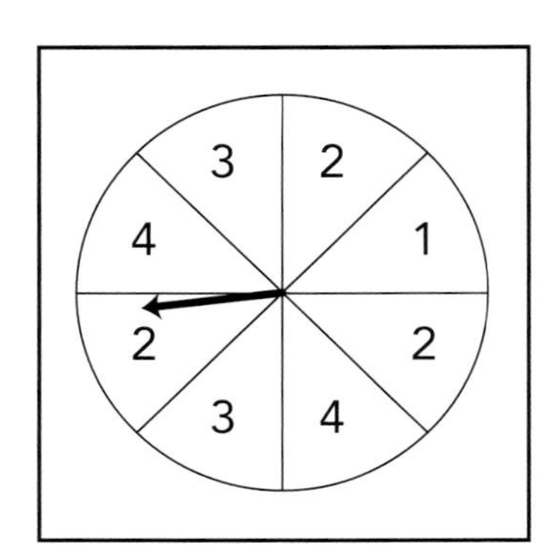

Find each probability

3.

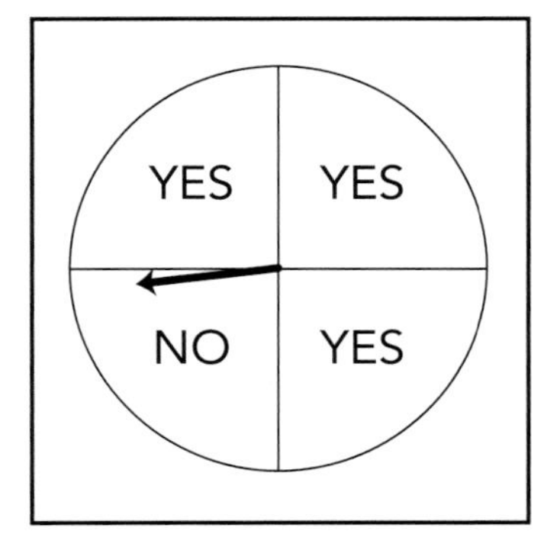

$P(\text{YES})$ $\frac{3}{4}$

$P(\text{NO})$ $\frac{1}{4}$

$P(\text{MAYBE})$ 0

4.

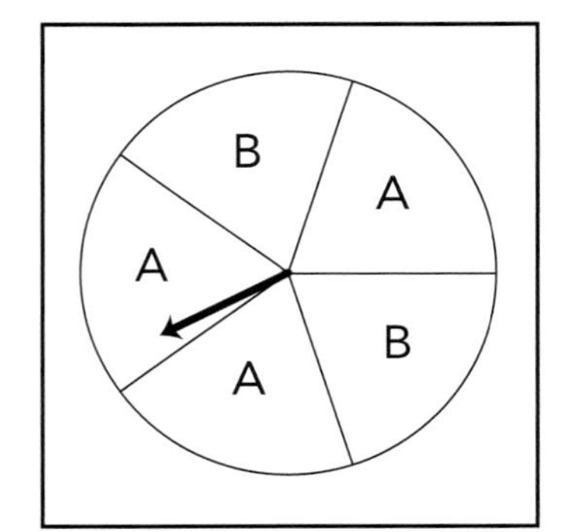

$P(A)$ $\frac{3}{5}$

$P(B)$ $\frac{2}{5}$

$P(A \text{ or } B)$ 1

Answer Key

Challenge! How do you use the number of sections in a spinner when finding the probability of an event?

Challenge: (Sample) The number of sections in the spinner is the denominator for the fraction when finding probability.

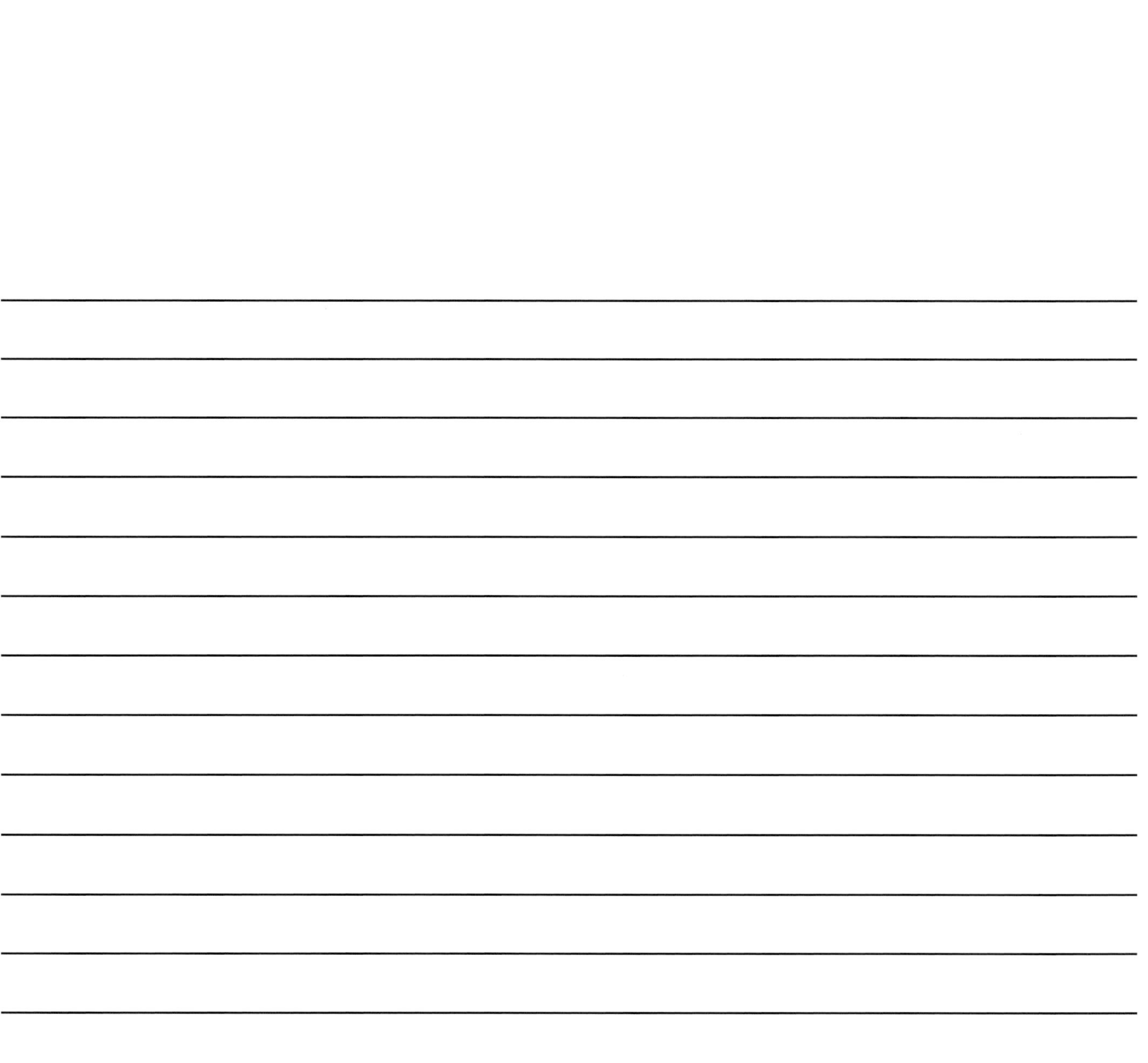

LESSON 4

Modeling Probability: Relationships Between Events

Objective

Model relationships between events using random drawings from a bag.

Common Core State Standards

- **7.SP.7a** Develop a uniform probability model by assigning equal probability to all outcomes, and use the model to determine probabilities of events. *For example, if a student is selected at random from a class, find the probability that Jane will be selected and the probability that a girl will be selected.*
- **7.SP.8a** Understand that, just as with simple events, the probability of a compound event is the fraction of outcomes in the sample space for which the compound event occurs.

Once students have a basic grasp of probability concepts, they can extend their understanding. They can reason, for example, about relationships between events, such as the probability that an event will occur and the probability it will not occur. Random drawings from a bag are a good way to model real relationships, such as between randomly selected people.

Try It! *Perform the Try It! activity on the next page.*

Talk About It

Discuss the Try It! activity.

- **Ask:** *What is the sum of the probability that an event will occur and the probability the event will not occur?*
- Reiterate the idea that picking a blue duck and picking a green duck cannot happen at the same time. Explain that this point is not as trivial as it seems. Ask students whether picking a blue tile and picking a square tile can happen at the same time. Mention to students that sometimes they will need to deal with questions like this.

Solve It

Reread the problem with students. Have them do the activities and explain how they determined the probabilities. Have them identify other probabilities that add up to 1.

More Ideas

For other ways to teach about relationships between events—

- Give spinners to pairs of students. Have students find probabilities such as the probability of spinning a 5 and the probability of not spinning a 5, or the probability of spinning a 4 or a 5.
- Have pairs of students use polyhedral dice to determine the probabilities, for example, of rolling and not rolling a multiple of 3 and the probability of rolling a prime number or a 4.

Formative Assessment

Have students try the following problem.

A bag contains 3 blue tiles, 4 green tiles, 1 yellow tile, and 2 red tiles. What is the probability of drawing a green or red tile from the bag?

A. $\frac{1}{5}$ **B.** $\frac{2}{5}$ **C.** $\frac{1}{2}$ **D.** $\frac{3}{5}$

Try It! 25 minutes | Groups of 4

Here is a problem about relationships between events.

Eve picks a prize from a bag filled with 3 blue ducks, 2 yellow ducks, 6 green ducks, and 1 red duck. What is the probability that Eve picks a yellow duck and what is the probability that she does not pick a yellow duck? What is the probability that Eve picks either a blue duck or a green duck?

Introduce the problem. Then have students do the activity to solve the problem. Distribute the materials.

Materials

- Color Tiles (6 blue, 4 yellow, 2 red, and 12 green per group)
- paper (1 sheet per group)

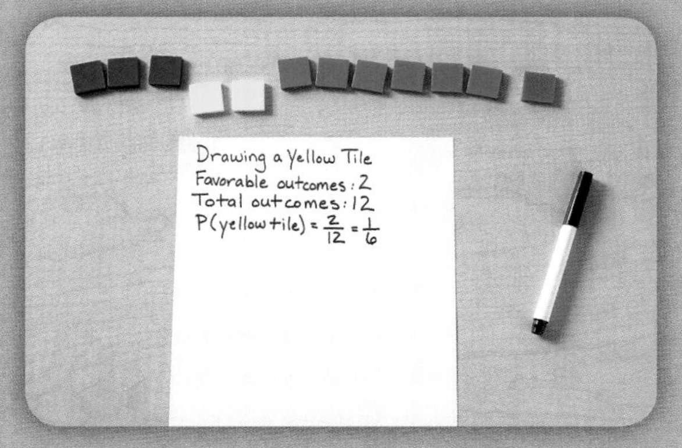

1. Say: *Use Color Tiles to model the ducks in the bag.* **Ask:** *What is the probability of drawing a yellow tile?* Have students model the possibilities and then count and record the favorable outcomes and the total outcomes. Have them express the probability as a fraction in simplest form.

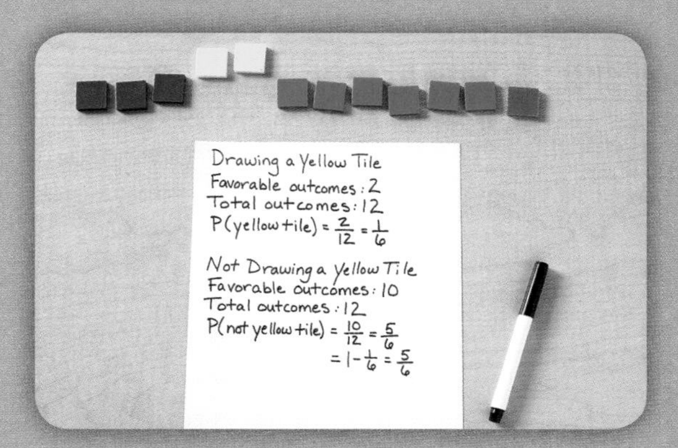

2. Ask: *What is the probability of drawing a tile that is not yellow?* Have students show the non-yellow tiles. **Say:** *Find and record the probability by listing and counting outcomes and then by subtracting from 1 the probability of drawing a yellow tile.*

3. Ask: *Can drawing a yellow tile happen at the same time as drawing a non-yellow tile? Can drawing a blue tile happen at the same time as drawing a green tile?* Have students model the events to show that both cannot occur at the same time.

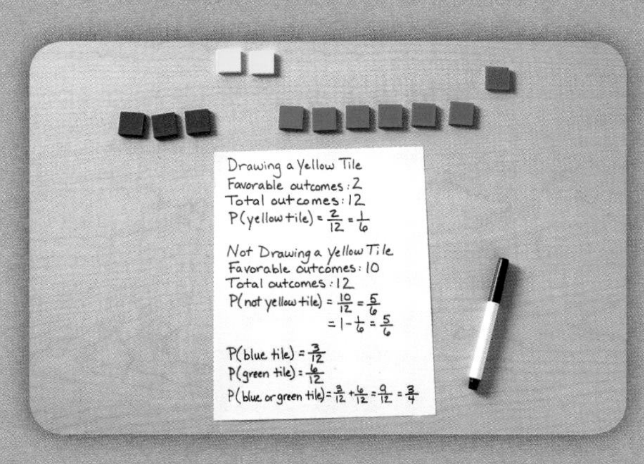

4. Ask: *What is the probability of drawing either a blue or a green tile?* Have students model the number of favorable selections—3 blue plus 6 green. Help them conclude that *P*(blue or green) is $\frac{9}{12}$, or $\frac{3}{4}$. Elicit, further, that this is the same as *P*(blue) + *P*(green)—that is, $\frac{3}{12} + \frac{6}{12} = \frac{9}{12} = \frac{3}{4}$.

Use Color Tiles to model a set with 3 yellow, 3 red, 4 blue, and 3 green. Find the probability of each event. (Check students' work.)

1.

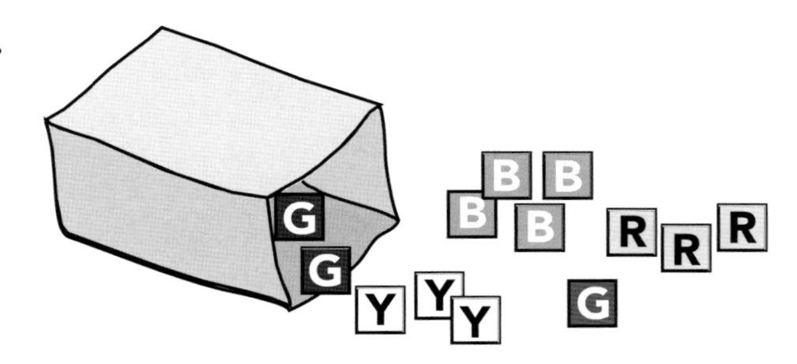

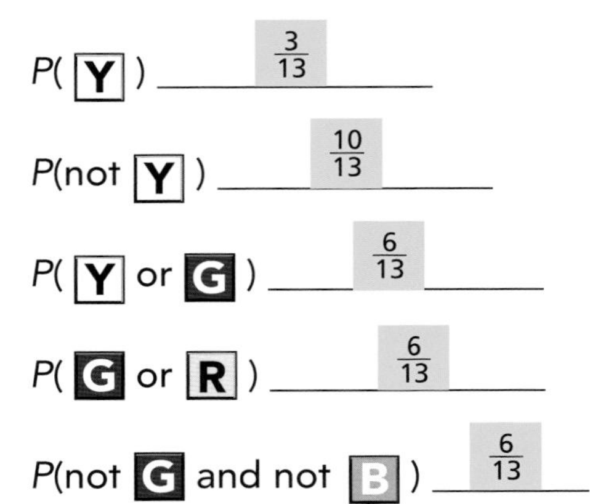

$P($Y$)$ $\frac{3}{13}$

$P($not Y$)$ $\frac{10}{13}$

$P($Y or G$)$ $\frac{6}{13}$

$P($G or R$)$ $\frac{6}{13}$

$P($not G and not B$)$ $\frac{6}{13}$

Using Color Tiles, model the set described. Then find the probability of each event.

2. Bag with 4 red tiles, 5 blue tiles, 6 green tiles, and 2 yellow tiles.

P(yellow or blue) $\frac{7}{17}$

P(red) $\frac{4}{17}$

P(green, red, or blue) $\frac{15}{17}$

3. Bag with 5 red tiles, 3 blue tiles, and 3 yellow tiles.

P(not blue and not red) $\frac{3}{11}$

P(yellow) $\frac{3}{11}$

P(red or yellow) $\frac{8}{11}$

Find each probability given the set described.

4. Bag with 10 red marbles, 12 blue marbles, 8 white marbles, 6 green marbles, and 4 yellow marbles.

P(not yellow) $\frac{9}{10}$

P(not red and not white) $\frac{11}{20}$

P(green or blue) $\frac{9}{20}$

P(not green) $\frac{17}{20}$

P(not green and not blue) $\frac{11}{20}$

P(green) $\frac{3}{20}$

5. Bag with 1 red marble, 1 blue marble, 1 white marble, 8 green marbles, and 10 yellow marbles.

P(blue) $\frac{1}{21}$

P(red) $\frac{1}{21}$

P(green) $\frac{8}{21}$

P(white) $\frac{1}{21}$

P(red, blue, white, green, or yellow) 1

P(not yellow) $\frac{11}{21}$

Challenge! If you have 20 items in a set and 4 of the items are red, what do you know about the probability of red and the probability of not red? Show your work.

Challenge: (Sample) The sum will be 1. Red is 4 out of 20 or $\frac{1}{5}$ and not red is 16 out of 20, or $\frac{4}{5}$; $\frac{1}{5} + \frac{4}{5} = 1$.

Objective

Determine whether a spinner is fair by comparing its fractional parts.

Common Core State Standards

- **7.SP.6** Approximate the probability of a chance event by collecting data on the chance process that produces it and observing its long-run relative frequency, and predict the approximate relative frequency given the probability. *For example, when rolling a number cube 600 times, predict that a 3 or 6 would be rolled roughly 200 times, but probably not exactly 200 times.*
- **7.SP.7b** Develop a probability model (which may not be uniform) by observing frequencies in data generated from a chance process. *For example, find the approximate probability that a spinning penny will land heads up or that a tossed paper cup will land open-end down. Do the outcomes for the spinning penny appear to be equally likely based on the observed frequencies?*

Statistics and Probability

Probability and Fairness

Using an area model to determine theoretical probability involves the understanding that the sum of the fractional parts of a whole must be 1. Using an area model also facilitates an understanding of "fairness." This activity uses dartboards as a starting point for the investigation of these concepts.

Try It! *Perform the Try It! activity on the next page.*

Talk About It

Discuss the Try It! activity.

- **Ask:** *What point values could you assign to each of the colored areas in order to make the second dartboard fair?*
- **Ask:** *In 20 tries, about how many times should you have hit red, blue, green, and yellow on the first dartboard? The second dartboard?*
- **Ask:** *What is the result if you add together the fractional parts of the first board? If you add together the fractional parts of the second board? In general, what should be the sum of the fractional parts of any board?*

Solve It

Reread the problem with students. Have students explain in writing how they determined whether the boards were fair or unfair.

More Ideas

For another way to teach about fair and unfair outcomes—

- Set up an experiment that will produce unfair outcomes. Place 2 yellow, 4 green, and 6 red Centimeter Cubes in a paper bag. Tell students not to look in the bag until the end of the activity. Have students pull out a cube, record its color, return the cube to the bag, and mix the cubes. Students should repeat the process 50 times. **Say:** *Examine the data from your experiment.* **Ask:** *What colors of cubes are there in the bag? Are the colors represented equally? If not, estimate the proportions that are represented.* When students have made and defended their predictions, have them open the bag and check their work.

Formative Assessment

Have students try the following problem.

Which of the following dartboards is unfair?

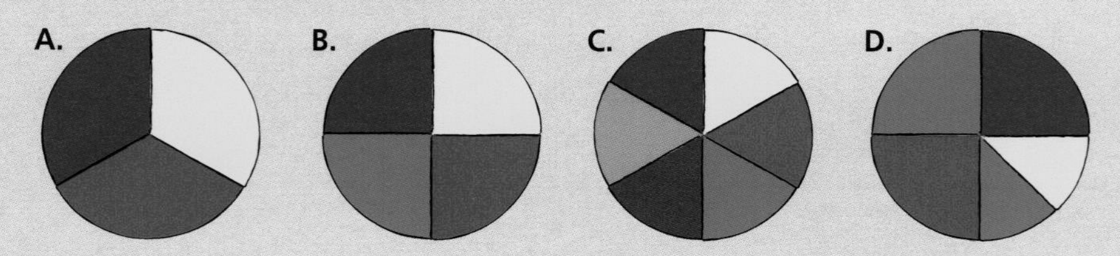

Try It! 30 minutes | Pairs

Here is a problem about determining whether a spinner is fair or unfair.

James and his friends play a magnetic dart game at their school's afternoon recreational program. Each player picks a different color as his or her target. The dartboard is mounted on a stand and spun. The players are blindfolded and take turns throwing darts at the spinning board, hoping to hit their target section. The players have two boards they can use. What is the probability of hitting each of the colors on the dartboards? Are the boards fair?

Introduce the problem. Then have students do the activity to solve the problem. Distribute the materials.

Materials
- Spinners

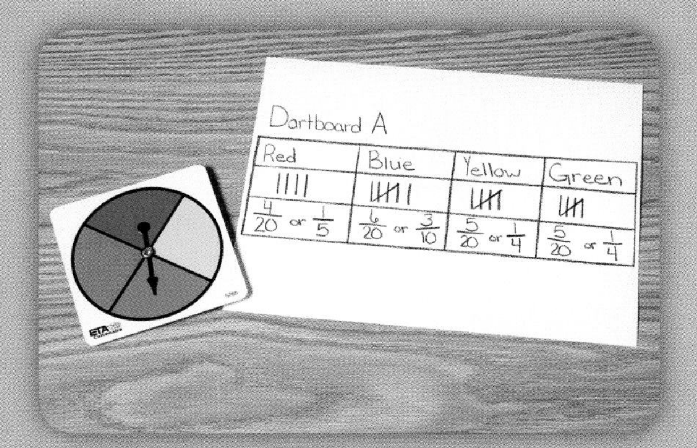

Dartboard A

Red	Blue	Yellow	Green
IIII	𝍸 I	𝍸	𝍸
$\frac{4}{20}$ or $\frac{1}{5}$	$\frac{6}{20}$ or $\frac{3}{10}$	$\frac{5}{20}$ or $\frac{1}{4}$	$\frac{5}{20}$ or $\frac{1}{4}$

1. Have students spin the spinner, which is divided into four equal parts, 20 times and record the results on a tally chart. **Ask:** *Based on your results, what is the probability of hitting each of the colors on this spinner?*

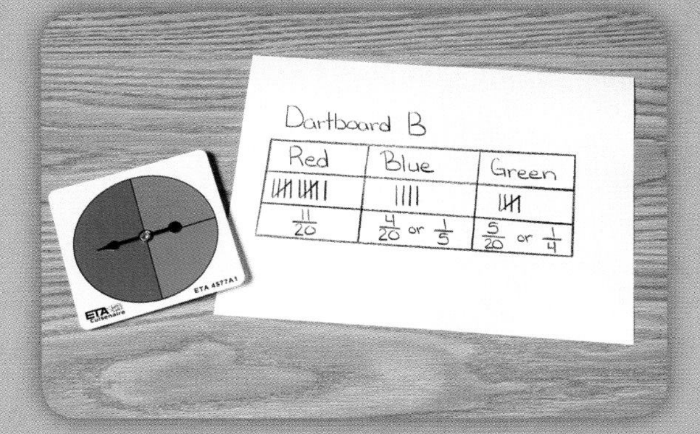

Dartboard B

Red	Blue	Green
𝍸 𝍸 I	IIII	𝍸
$\frac{11}{20}$	$\frac{4}{20}$ or $\frac{1}{5}$	$\frac{5}{20}$ or $\frac{1}{4}$

2. Now have students repeat the experiment with the spinner that is divided into one half and two quarters. **Ask:** *Based on your results, what is the probability of hitting each of the colors on this spinner?*

Dartboard A

Red	Blue	Yellow	Green
IIII	𝍸 I	𝍸	𝍸
$\frac{4}{20}$ or $\frac{1}{5}$	$\frac{6}{20}$ or $\frac{3}{10}$	$\frac{5}{20}$ or $\frac{1}{4}$	$\frac{5}{20}$ or $\frac{1}{4}$

P(Red) = $\frac{1}{4}$ P(Blue) = $\frac{1}{4}$ P(Yellow) = $\frac{1}{4}$ P(Green) = $\frac{1}{4}$

Dartboard B

Red	Blue	Green
𝍸 𝍸 I	IIII	𝍸
$\frac{11}{20}$	$\frac{4}{20}$ or $\frac{1}{5}$	$\frac{5}{20}$ or $\frac{1}{4}$

P(Red) = $\frac{1}{2}$ P(Blue) = $\frac{1}{4}$ P(Green) = $\frac{1}{4}$

3. Have students compare the results of their experiments to the actual probabilities of hitting each color. Ask them to summarize their findings. **Ask:** *Are both of the dartboards fair? Explain.*

⚠ Look Out!

Some students may think that if an outcome is possible, that it is as likely to occur as any other possible outcome. Help students realize that the area a section covers influences the probability that the spinner will land there.

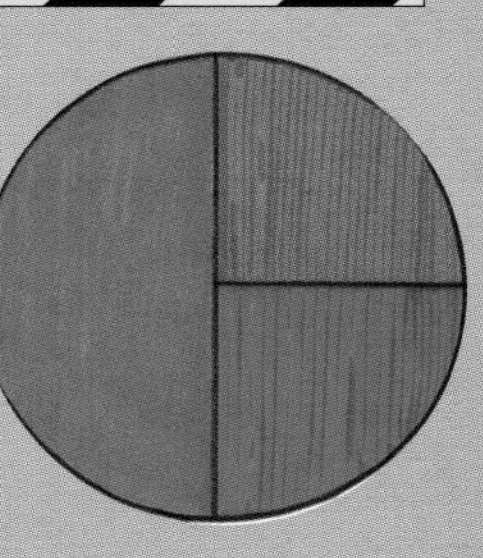

Use the spinner collection to model fair and unfair spinners. Find a spinner whose sections match each spinner below. Answer the questions.

1.

1 2 5 3 4

Find $P(1)$. $\frac{1}{5}$

Find $P(3)$. $\frac{1}{5}$

Find $P(4)$. $\frac{1}{5}$

Is the spinner fair? yes

Why or why not? all sections have the same probability

2.

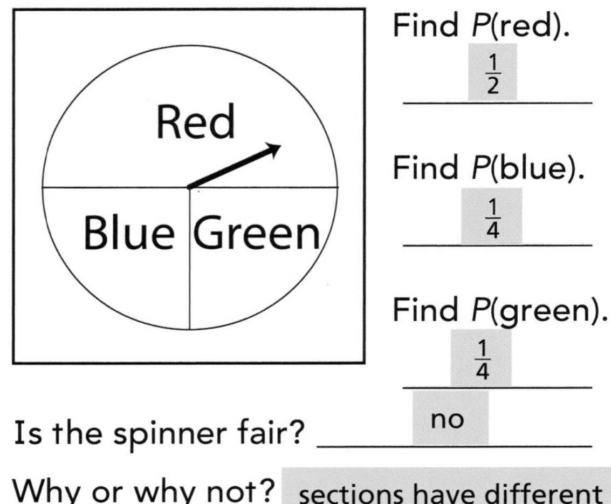

Find $P(\text{red})$. $\frac{1}{2}$

Find $P(\text{blue})$. $\frac{1}{4}$

Find $P(\text{green})$. $\frac{1}{4}$

Is the spinner fair? no

Why or why not? sections have different probabilities

Using the spinner collection, model a fair and an unfair spinner. Sketch the models. Answer the questions.

3. Sketch a fair spinner below.

Check students' models and explanations.

Why is the spinner fair? ______

4. Sketch an unfair spinner below.

Check students' models and explanations.

Why is the spinner unfair? ______

Determine if each spinner is fair. Explain your answer.

5.

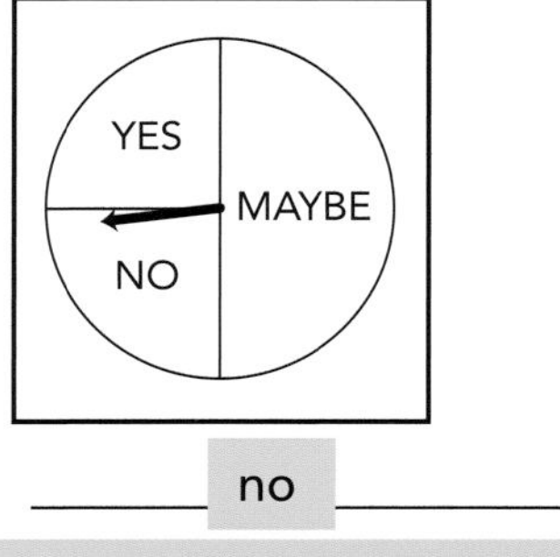

no

the outcomes have different probabilities

6.

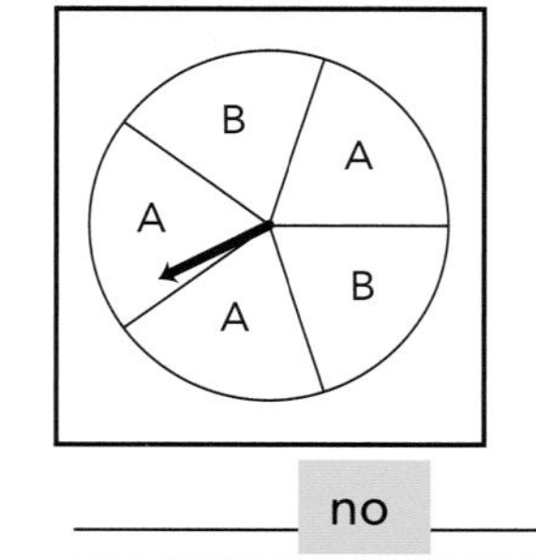

no

the probability is greater for spinning A

7.

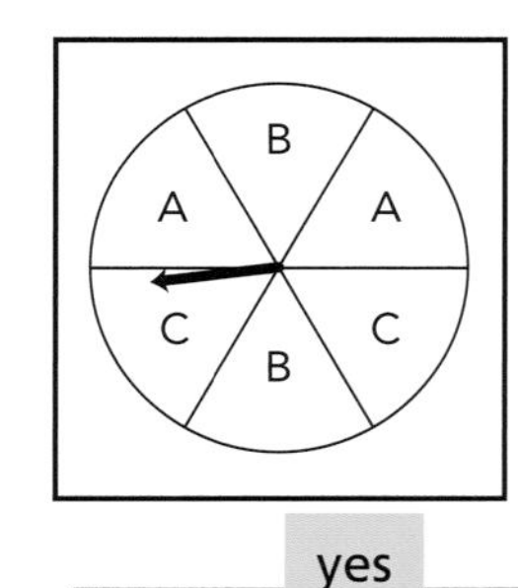

yes

the probability for each outcome is the same

Answer Key

Challenge! When a spinner has an odd number of equal-sized sections and the sections are not uniquely labeled, how can you be certain that the spinner is not fair? Are there any odd numbers for which the spinner could be fair? Explain or draw an example.

Challenge: (Sample) The number and size of the sections with different labels have to be equal. If a spinner has 5 sections, you can meet that requirement if none of the sections has the same label. A spinner with 9 equal sections could be fair if the sections were marked with 3 different labels equally. This would be true for a spinner with 15 sections and 5 different labels. This type of pattern is true for spinners that have a number of labels that is a factor of the number of sections.

Objective

Determine the probability for a random drawing without replacement.

Common Core State Standards

- **7.SP.6** Approximate the probability of a chance event by collecting data on the chance process that produces it and observing its long-run relative frequency, and predict the approximate relative frequency given the probability. *For example, when rolling a number cube 600 times, predict that a 3 or 6 would be rolled roughly 200 times, but probably not exactly 200 times.*
- **7.SP.8a** Understand that, just as with simple events, the probability of a compound event is the fraction of outcomes in the sample space for which the compound event occurs.
- **7.SP.8b** Represent sample spaces for compound events using methods such as organized lists, tables and tree diagrams. For an event described in everyday language (e.g., "rolling double sixes"), identify the outcomes in the sample space which compose the event.

Statistics and Probability

Finding Probability Without Replacement

Students have learned how to find the probability of a single event. In this lesson, students will determine the probability of having the same outcome occur more than once in a row.

Try It! *Perform the Try It! activity on the next page.*

Talk About It

Discuss the Try It! activity.

- **Say:** *The probability of picking a blue sock can be found by taking the number of blue socks and dividing that number by the total number of socks in the drawer.* **Ask:** *What is the probability of picking a blue sock?* Elicit that the probability is $\frac{2}{6}$ or $\frac{1}{3}$.
- **Say:** *Suppose that Bill picks a blue sock and doesn't return it to the drawer.* **Ask:** *If he picks again, what are his chances of picking another blue sock?* Elicit that because there are now just five socks in the drawer and only one of them is blue, you divide one by five; the answer is $\frac{1}{5}$.
- **Ask:** *How could you use your data from the first step to estimate the probability of drawing two blue socks?* Have students obtain an estimate and compare it with the theoretical probability.

Solve It

Reread the problem with students. Ask them to write to Bill explaining how to determine the probability of picking two blue socks.

More Ideas

For another way to teach about probability for a random drawing—

- Extend the activity by having students determine the probability of picking two blue socks with replacement. Have students re-enact the scenario, but this time they should replace the first sock drawn. Compare the probabilities (both theoretical and experimental) with and without replacement.

Formative Assessment

Have students try the following problem.

Each of the numbers 1–6 is written on a card and put into a hat. Two cards will be drawn, one at a time, without replacement. What is the probability of drawing two odd numbers?

A. $\frac{7}{10}$ **B.** $\frac{1}{3}$ **C.** $\frac{1}{5}$ **D.** $\frac{1}{10}$

Try It! **40 minutes | Pairs**

Here is a problem about a random drawing without replacement.

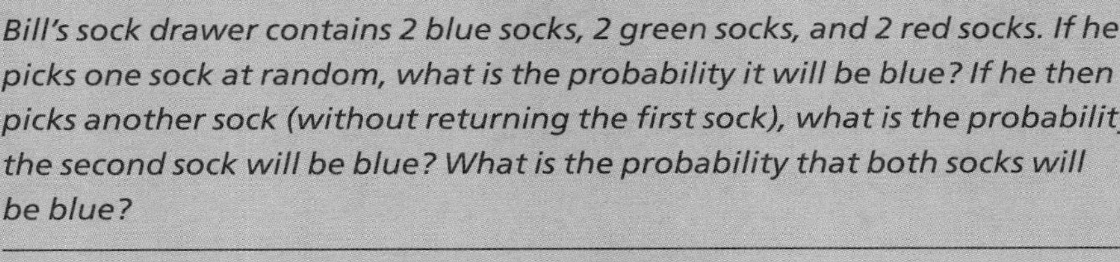

Bill's sock drawer contains 2 blue socks, 2 green socks, and 2 red socks. If he picks one sock at random, what is the probability it will be blue? If he then picks another sock (without returning the first sock), what is the probability the second sock will be blue? What is the probability that both socks will be blue?

Introduce the problem. Then have students do the activity to solve the problem. Distribute the materials.

Materials

- Centimeter Cubes (2 blue, 2 green, and 2 red)
- BLM 12
- paper bag

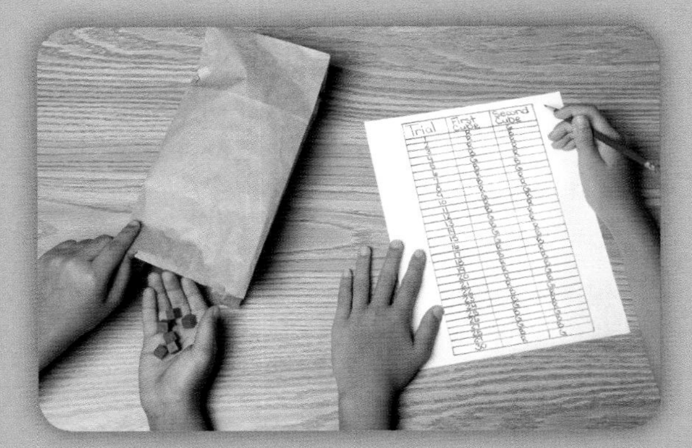

1. Have students place 2 blue, 2 green, and 2 red cubes in a paper bag. One student should pick a cube at random. The other student should record its color on the recording chart. Without returning the cube to the bag, the first student should select another cube at random. Its color also should be recorded on the chart. Students should return the cubes to the bag and repeat the procedure 30 times.

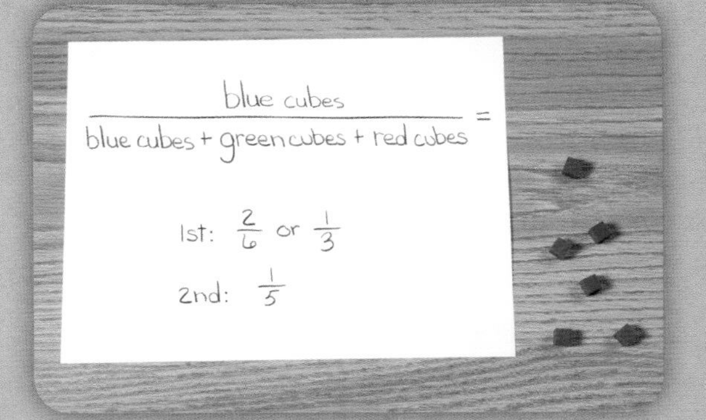

2. Have students determine the probability of randomly choosing a blue sock from the sock drawer. Then have them determine the probability that the second sock drawn will be blue if the first sock drawn is blue and is not replaced.

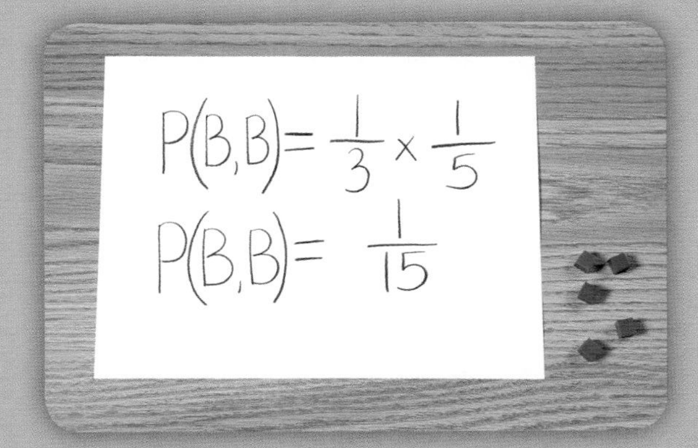

3. Have students use the formula to determine the probability that both socks Bill draws will be blue. Remind students of the Conditional Probability formula, $P(A \text{ and } B) = P(A) \times P(B \text{ given } A)$.

⚠ Look Out!

Watch for students who do not decrease the number of socks available when Bill draws the second sock. Have students re-enact the scenario with cubes to confirm that the number of socks in the drawer is now five and not six.

Use Centimeter Cubes to model the probability of each event, without replacement. Make a bag like the one shown. Answer the questions. (Check students' work.)

1. The bag has 2 yellow cubes, 5 green cubes, and 3 red cubes.

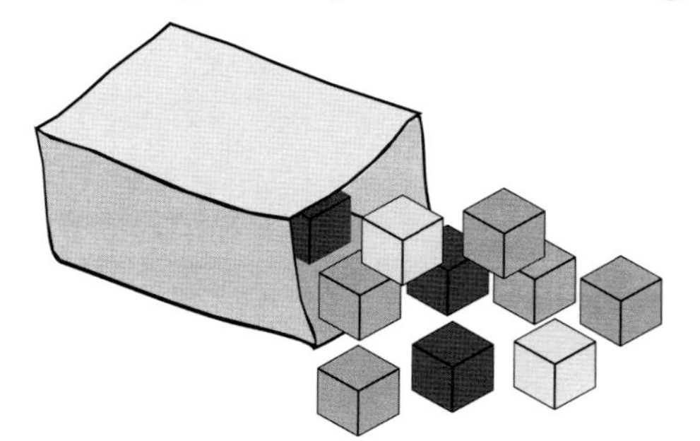

What is the probability of selecting a yellow cube at random? $\frac{1}{5}$

Without replacing the yellow cube, what is the probability of selecting a red cube at random? $\frac{1}{3}$

What is P(yellow, red)? $\frac{1}{15}$

What is P(yellow, yellow)? $\frac{1}{45}$

Using Centimeter Cubes, model each bag described. Find each probability without replacement.

2. A bag with 5 black cubes, 3 pink cubes, and 2 blue cubes

   What is P(blue, black)? $\frac{1}{9}$

   What is P(pink, blue)? $\frac{1}{15}$

   What is P(black, black)? $\frac{2}{9}$

3. A bag with 6 orange cubes, 6 red cubes, and 6 brown cubes

   What is P(orange, red)? $\frac{2}{17}$

   What is P(red, red)? $\frac{5}{51}$

   What is P(brown, red)? $\frac{2}{17}$

Find each probability without replacement.

4. A bag with 5 black marbles, 2 white marbles, and 8 yellow marbles

   What is P(yellow, white)? $\frac{8}{105}$

   What is P(white, black)? $\frac{1}{21}$

   What is P(black, black)? $\frac{2}{21}$

   What is P(black, yellow)? $\frac{4}{21}$

5. A bag with 3 solid ribbons, 4 striped ribbons, and 10 checkered ribbons

   What is P(solid, solid)? $\frac{3}{136}$

   What is P(checkered, striped)? $\frac{5}{34}$

   What is P(striped, solid)? $\frac{3}{68}$

   What is P(solid, checkered)? $\frac{15}{136}$

6. A bag with 12 red tiles, 10 black tiles, and 20 white tiles

   What is P(red, white)? $\frac{40}{287}$

   What is P(white, black)? $\frac{100}{861}$

   What is P(red, black)? $\frac{20}{287}$

   What is P(black, black)? $\frac{15}{287}$

7. A bag with 10 green marbles, 2 clear marbles, and 8 blue marbles

   What is P(clear, clear)? $\frac{1}{190}$

   What is P(green, clear)? $\frac{1}{19}$

   What is P(blue, green)? $\frac{4}{19}$

   What is P(green, green)? $\frac{9}{38}$

Answer Key

Challenge! Describe the numbers you multiply in the denominator when you find the probability of two events without replacement. When does the probability in simplest form have a denominator that differs from the product of the numbers you just described?

Challenge: (Sample) The denominator is the product of the number of items and one less than the number of items. When a numerator and a denominator have a common factor and you divide each by the common factor, the denominator of the probability in simplest form is not the product of the number of items times one less than the number of items.

Objective

Find the theoretical and experimental probabilities of an event involving dice.

Common Core State Standards

- **7.SP.6** Approximate the probability of a chance event by collecting data on the chance process that produces it and observing its long-run relative frequency, and predict the approximate relative frequency given the probability. *For example, when rolling a number cube 600 times, predict that a 3 or 6 would be rolled roughly 200 times, but probably not exactly 200 times.*
- **7.SP.8a** Understand that, just as with simple events, the probability of a compound event is the fraction of outcomes in the sample space for which the compound event occurs.
- **7.SP.8b** Represent sample spaces for compound events using methods such as organized lists, tables and tree diagrams. For an event described in everyday language (e.g., "rolling double sixes"), identify the outcomes in the sample space which compose the event.

Statistics and Probability

Theoretical and Experimental Probability with Dice

Students' experiences with determining theoretical probability continue with this activity in which they play a game involving dice. The students then gather experimental data by playing the game and comparing the results with the mathematically determined theoretical probabilities.

Try It! *Perform the Try It! activity on the next page.*

Talk About It

Discuss the Try It! activity.

- **Ask:** *Based on your experiment, what is* P(*multiple of 3) and* P(*multiple of 4)? How do these values compare with the theoretical probabilities?* Have students share their results.

Solve It

Reread the problem with students. Discuss the differences between the experimental and theoretical probabilities. Ask students what they would do to obtain experimental probabilities closer to the theoretical probabilities. Elicit that they could increase the number of rolls.

More Ideas

For another way to teach about theoretical and experimental probabilities—

- Have students complete a similar activity using the 4-sided die and the 8-sided die. This time have them determine *P*(sum of 4) and *P*(sum of 5). Tell students to find the experimental probability by rolling the dice 32 times and the theoretical probability using a table of values.

Formative Assessment

Have students try the following problem.

A green 6-sided number cube and a blue 6-sided number cube are rolled, and the results are added together. All of the possible sums are listed in the table.

Which of the following probabilities is correct?

A. *P*(multiple of 10) = $\frac{1}{4}$

B. *P*(odd number) = $\frac{1}{2}$

C. *P*(even number) = $\frac{1}{36}$

D. *P*(greater than 5) = $\frac{1}{4}$

Blue \ Green	1	2	3	4	5	6
1	2	3	4	5	6	7
2	3	4	5	6	7	8
3	4	5	6	7	8	9
4	5	6	7	8	9	10
5	6	7	8	9	10	11
6	7	8	9	10	11	12

Try It! 25 minutes | Pairs

Here is a problem about theoretical probability vs. experimental probability.

Matt and Jana are playing a game. Matt rolls a 4-sided die, and Jana rolls an 8-sided die. If the product of the rolls is a multiple of 3, Matt gets a point. If the product of the rolls is a multiple of 4, Jana gets a point. A game consists of 32 rolls of the dice. Who has a better chance of winning?

Introduce the problem. Then have students do the activity to solve the problem. Distribute the materials.

Materials
- Polyhedral Dice Set (4-sided die and 8-sided die)

1. Have students take the roles of Matt and Jana and play a round (32 rolls of the dice) of the game. Students should record the product for each of the rolls. **Ask:** *Who won the game—Matt or Jana?*

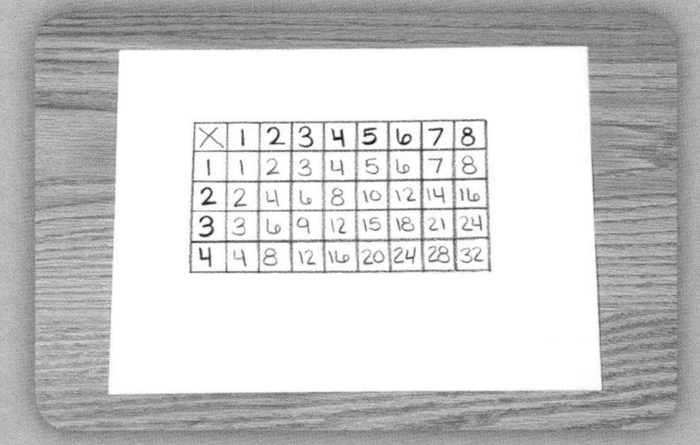

×	1	2	3	4	5	6	7	8
1	1	2	3	4	5	6	7	8
2	2	4	6	8	10	12	14	16
3	3	6	9	12	15	18	21	24
4	4	8	12	16	20	24	28	32

2. Say: *Now determine the theoretical probability.* Have students set up a table of values and fill in the products.

×	1	2	3	4	5	6	7	8
1	1	2	3	4	5	6	7	8
2	2	4	6	8	10	12	14	16
3	3	6	9	12	15	18	21	24
4	4	8	12	16	20	24	28	32

3. Have students mark the multiples of 3 and the multiples of 4. **Ask:** *How many values in the table are multiples of 3? How many are multiples of 4?* Elicit from students that there are fourteen multiples of 3 and sixteen multiples of 4.

×	1	2	3	4	5	6	7	8
1	1	2	3	4	5	6	7	8
2	2	4	6	8	10	12	14	16
3	3	6	9	12	15	18	21	24
4	4	8	12	16	20	24	28	32

$P(\text{multiple of } 3) = \frac{14}{32} = \frac{7}{16}$

$P(\text{multiple of } 4) = \frac{16}{32} = \frac{1}{2}$

4. Have students then determine *P*(multiple of 4) and *P*(multiple of 3) using their table of values. **Ask:** *Who has a better chance of winning a round of this game?*

Use a 4-sided die and a 6-sided die to make a table of products when the dice are rolled. Use the table to find each probability.

(Check students' work.)

1.

×	1	2	3	4	5	6
1	1	2	3	4	5	6
2	2	4	6	8	10	12
3	3	6	9	12	15	18
4	4	8	12	16	20	24

P(multiple of 6) $\frac{1}{3}$

P(multiple of 4) $\frac{11}{24}$

P(even product) $\frac{3}{4}$

P(multiple of 10) $\frac{1}{12}$

Using Polyhedral Dice, make a table to find each probability.

2. two 4-sided dice

Check students' tables.

P(multiple of 3) $\frac{7}{16}$

P(product < 15) $\frac{15}{16}$

P(product that is a prime number) $\frac{1}{4}$

P(multiple of 8) $\frac{3}{16}$

3. 6-sided die and 10-sided die

Check students' tables.

P(odd product) $\frac{1}{4}$

P(product < 10) $\frac{1}{3}$

P(product > 40) $\frac{1}{10}$

P(multiple of 5) $\frac{1}{3}$

Answer Key

Challenge! An experiment has you roll an 8-sided die and a 12-sided die and multiply the face values of the dice. What is the number of outcomes for this experiment? What is the smallest product in the table? What is the largest product in the table? How many products are less than 10?

Challenge: 96; 1; 96; 25

Objective

Find the probability of a compound event; make an organized list.

Common Core State Standards

- **7.SP.6** Approximate the probability of a chance event by collecting data on the chance process that produces it and observing its long-run relative frequency, and predict the approximate relative frequency given the probability. *For example, when rolling a number cube 600 times, predict that a 3 or 6 would be rolled roughly 200 times, but probably not exactly 200 times.*
- **7.SP.8a** Understand that, just as with simple events, the probability of a compound event is the fraction of outcomes in the sample space for which the compound event occurs.
- **7.SP.8b** Represent sample spaces for compound events using methods such as organized lists, tables and tree diagrams. For an event described in everyday language (e.g., "rolling double sixes"), identify the outcomes in the sample space which compose the event.

Compound Events: Making an Organized List

Experience with experiments helps students build on their intuitive sense about probability. In this lesson, students make an organized list to identify outcomes in a sample space and make predictions about their occurrences. Comparing theoretical predictions and observational data enables students to draw new insights and adjust their thinking accordingly.

Try It! *Perform the Try It! activity on the next page.*

Talk About It

Discuss the Try It! activity.

- **Ask:** *What fraction do you use to compute the probability of an event?* Point out that the fraction is the number of favorable outcomes over the number of possible outcomes.
- **Ask:** *Besides using an organized list, how else might you find the number of possible outcomes?* If appropriate, elicit that the possibilities include drawing a picture and making a tree diagram.

Solve It

Reread the problem with students. Have them find and write the probability for each single event in the form of a fraction. Guide students to multiply these fractions to find the probability of the compound event. Have students compare the experimental and theoretical probabilities. Ask students to tell how the favorable outcomes would differ if Deana wanted to roll a number greater than 5 OR toss yellow.

More Ideas

For other ways to teach about probabilities of compound events—

- Have students conduct the experiment using a 1–8 spinner instead of the die.
- Use Color Tiles. Place six yellow and four red tiles in a bag. Have students draw one tile, record the color, replace the tile, and repeat. Each draw is a single event. The two draws are a combined event. Have students find the experimental and theoretical probabilities of drawing a yellow tile first and a red tile second.

Formative Assessment

Have students try the following problem.

Johann rolls a number cube with the numbers 1–6. He also tosses a coin. What is the probability that he will roll an even number and toss tails?

A. $\frac{1}{2}$ **B.** $\frac{1}{3}$ **C.** $\frac{1}{4}$ **D.** $\frac{1}{12}$

Try It! 30 minutes | Groups of 4

Here is a problem about the probability of a compound event.

Deana has a polyhedral die with faces labeled 1–8 and a counter with one yellow and one red face. What is the probability that she will roll a number greater than 5 and toss a counter yellow-face up?

Introduce the problem. Then have students do the activity to solve the problem. Distribute the materials.

Materials

- Octahedral Dice (1 per group)
- Two-Color Counters (1 per group)
- cup (optional for rolling die and counters; 1 per group)
- paper (2 sheets per group)

1. Ask: *What are the favorable outcomes for each single event—that is, for just the number and for just the color? What are the favorable outcomes for the compound event?* Have students record the favorable outcomes for the compound event.

2. Guide students to list all the possible outcomes for the compound event. Then have them perform at least 50 trials for this event and record their results. **Ask:** *What are the experimental and theoretical probabilities for the compound event?*

3. Guide students to see that they can determine the number of possible and favorable outcomes by using the Counting Principle. Show how this leads to the rule $P(A \text{ and } B) = P(A) \times P(B)$.

⚠ Look Out!

Emphasize that both events must have favorable outcomes to satisfy the conditions: rolling a 2 and yellow is not a favorable outcome because the outcome is not favorable for one of the two single events. Students may think that the theoretical probability and the experimental probability should be the same. Stress that the two results may not be the same.

Use the decahedral die and a Two-Color Counter to model each probability. Find the probability of each compound event. (Check students' work.)

1. 10-sided die numbered 0 to 9 and 1 Two-Color Counter

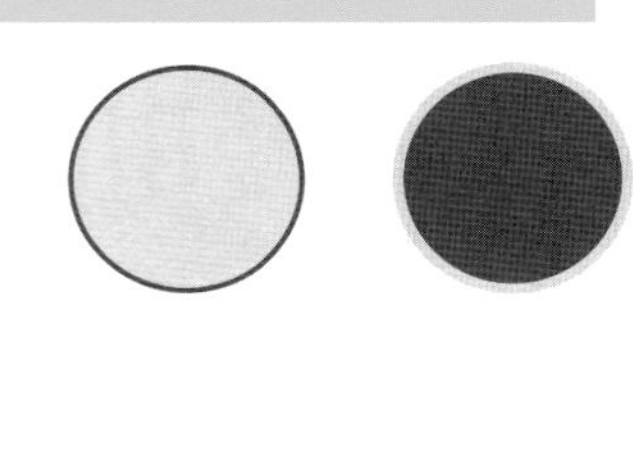

P(1 and red) $\frac{1}{20}$

P(8 and red) $\frac{1}{20}$

P(4 and not yellow) $\frac{1}{20}$

P(6 and yellow) $\frac{1}{20}$

P(7 or 8 and red) $\frac{1}{10}$

Using a die and a Two-Color Counter, model each probability. Find each probability.

2. 20-sided die numbered 1 to 20 and 1 counter

P(1 and yellow) $\frac{1}{40}$

P(12 and red) $\frac{1}{40}$

P(4, not red) $\frac{1}{40}$

3. 6-sided die numbered 1 to 6 and 1 counter

P(2 and red or yellow) $\frac{1}{6}$

P(not 3, red) $\frac{5}{12}$

P(2 and yellow) $\frac{1}{12}$

P(not 4 or 5, yellow) $\frac{1}{3}$

Find each probability.

4. 8-sided die numbered 1 to 8 and 1 counter

P(1 and yellow) $\frac{1}{16}$

P(7, not red) $\frac{1}{16}$

P(not 9, not yellow) $\frac{1}{2}$

P(5 or 6, red) $\frac{1}{8}$

5. 12-sided die numbered 1 to 12 and 1 counter

P(12 and yellow) $\frac{1}{24}$

P(13 and red) $\frac{1}{2}$

P(not 1, not yellow) $\frac{11}{24}$

P(4 and red or yellow) $\frac{1}{12}$

Answer Key

Challenge! What does the word *compound* mean when finding the probability of an event?

Challenge: (Sample) A compound event is an event with two parts that are independent of each other.

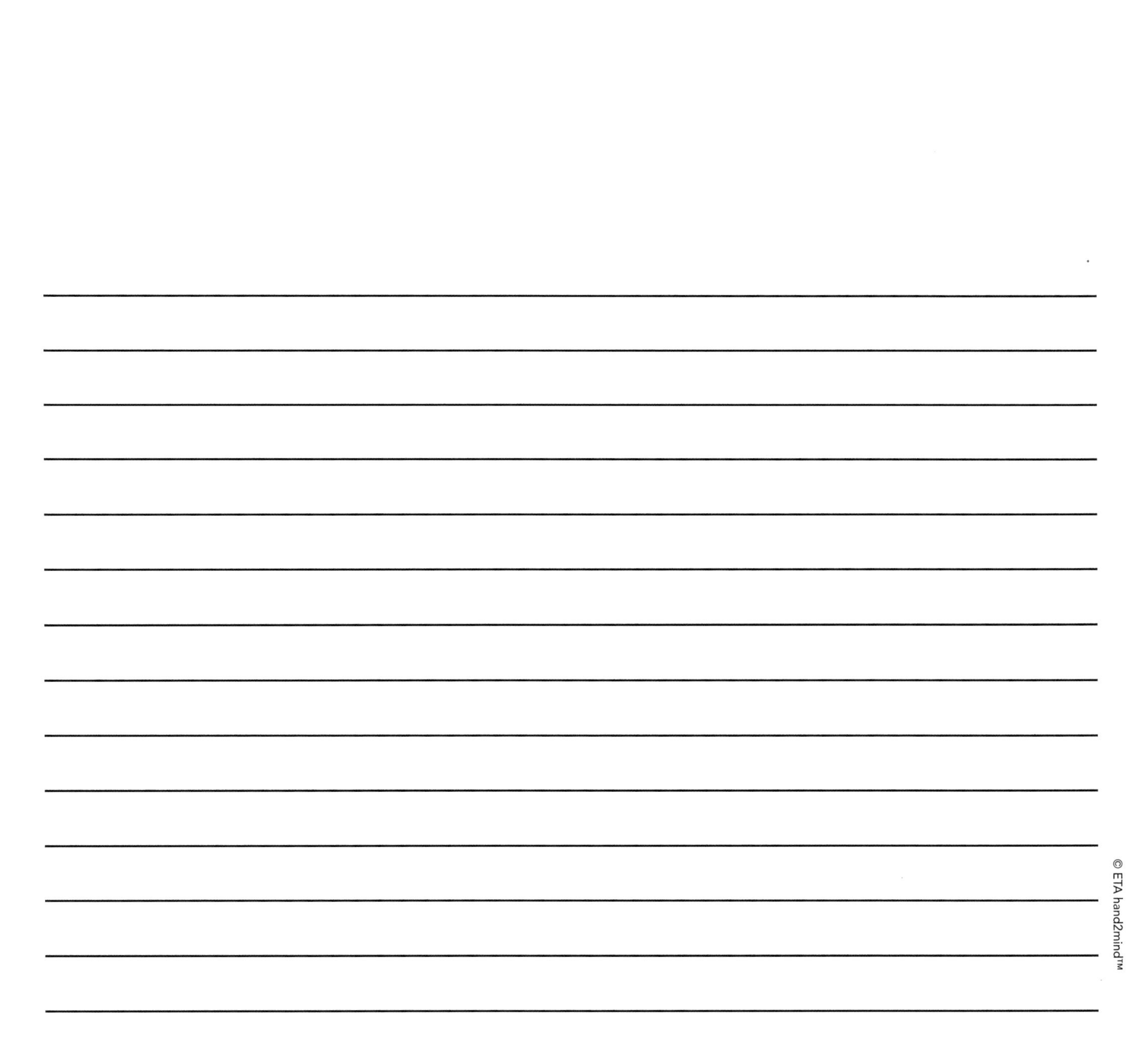

LESSON 9

Objective

Find the probability of a compound event; make a tree diagram.

Common Core State Standards

- **7.SP.8a** Understand that, just as with simple events, the probability of a compound event is the fraction of outcomes in the sample space for which the compound event occurs.
- **7.SP.8b** Represent sample spaces for compound events using methods such as organized lists, tables and tree diagrams. For an event described in everyday language (e.g., "rolling double sixes"), identify the outcomes in the sample space which compose the event.

Statistics and Probability

Compound Events: Making a Tree Diagram

Students have probably noticed a discrepancy between the theoretical probability of a simple event occurring and their actual experimental results. They should understand by now that generally, the more trials one performs, the closer the experimental results will be to the theoretical outcome. This activity lays the foundation for ensuring that the theoretical probability for compound events is correctly calculated.

Try It! *Perform the Try It! activity on the next page.*

Talk About It

Discuss the Try It! activity.

- **Ask:** *How many different combinations of colors and numbers do you think there are for the spinner and the die?*
- **Ask:** *What are the possible outcomes for a spin of the spinner? What are the possible outcomes for a roll of the die? Does the outcome of Jackie's spin affect the outcome of Jeff's roll of the die? Does the outcome of Jeff's roll of the die affect the outcome of Jackie's spin?*

Solve It

Have students explain in writing how likely Jeff and Jackie are to spin and roll a red 4. Are their chances of spinning a red 4 different from their chances of spinning and rolling any other combination?

More Ideas

For other ways to teach about probabilities of compound events—

- Have students extend this activity by finding the probabilities of other outcomes such as *P*(red, 3), *P*(blue, 1), *P*(green, 3), and so forth. Gradually increase the complexity to outcomes such as *P*((red, 4) *or* (red, 2)), *P*((green, 3) *or* (blue, 2)), and *P*((yellow, 3) *and* (blue, 4)).
- Have students conduct a similar experiment using two number cubes. Students can find the probability of rolling an odd number, then an even number.

Formative Assessment

Have students try the following problem.

A family has three children. What is the probability that two of the children are boys?

A. $\frac{1}{4}$ **B.** $\frac{1}{3}$ **C.** $\frac{3}{8}$ **D.** $\frac{1}{2}$

Try It! 30 minutes | Pairs

Here is a problem about the probability of a compound event.

Jeff has a spinner that is divided into four equal sections: red, yellow, green, and blue. Jackie has a 6-sided die. They decide to make up a game of their own in which Jeff must call out a number and Jackie must call out a color. If they call out "red" and "4," what is the probability that they will spin and roll that particular outcome?

Introduce the problem. Then have students do the activity to solve the problem. Distribute the materials.

Materials
- Spinners
- Number Cube

1. Have students simulate the game by spinning and rolling until they get "red" and "4." Have them record how many trials it took them to get that result. **Ask:** *How many trials did it take?*

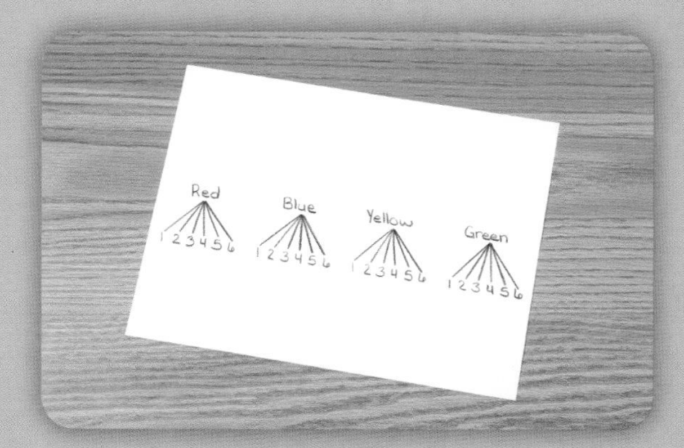

2. Next, have students create a tree diagram, starting with the spinner colors. Then have them complete the tree diagram with the numbers on the die. **Ask:** *What is the probability that you will get "red" and "4" on your first try?*

3. Discuss the tree diagram with students. **Ask:** *How many spins and rolls should it take, on average, to get "red" and "4"?*

Look Out!

Students may want to stop drawing the tree diagram once they have reached the outcome (red, 4). They need to understand that in order to find the total number of possible outcomes, they will need to complete the entire tree diagram. Explain to them that the total provides them with the correct denominator for the fraction that represents *P*(red, 4).

Use the 4-section color spinner and a number cube to simulate a game. Make a tree diagram for all possible outcomes. Find each probability.

(Check students' work.)

1. Four-section spinner with red, blue, green, and yellow sections and a number cube labeled 1 to 6

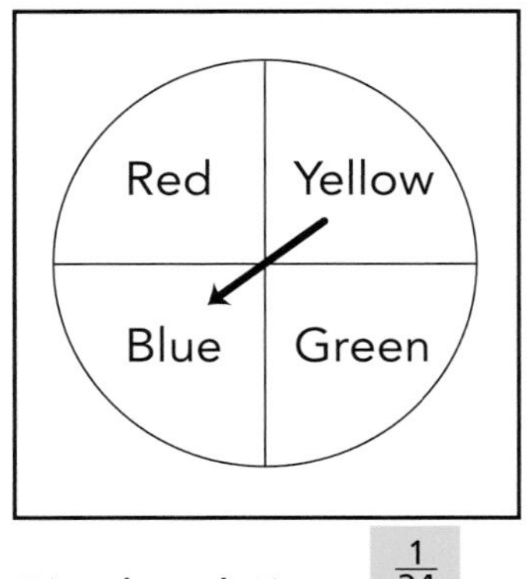

Check students' tree diagram. The diagram should have four branches, one for each color and each branch should have six numbers, 1–6.

P(red and 1) $\frac{1}{24}$

P(green and an even number) $\frac{1}{8}$

Using the 6-section color spinner and a coin, make a tree diagram of all possible outcomes. Find each probability.

2.

Check students' tree diagram. The diagram should have six branches, one for each color and each branch should have 2 options, heads and tails.

P(yellow and heads) $\frac{1}{12}$

P(blue or green and heads) $\frac{1}{6}$

Find each probability given the two elements of chance.

3. 2 coins

P(two heads) $\frac{1}{4}$

P(heads and tails) $\frac{1}{2}$

4. 8-sided die labeled 1–8 and a coin

P(8 and heads) $\frac{1}{16}$

P(tails and odd) $\frac{1}{4}$

5. two-section spinner labeled red and blue, and a 4-sided die labeled 1–4

P(blue and 3) $\frac{1}{8}$

P(red and even) $\frac{1}{4}$

Answer Key

Challenge! Describe a tree diagram for three items of chance: coin, 4-section spinner, and a number cube. Does the number of possible outcomes vary depending on the order in which you make your diagram? Explain.

Challenge: (Sample) The diagram could be in any order of the items of chance. If the diagram is made in the order the items are named, there will be two branches, one heads and one tails. Each of these branches will have four branches, one for each color. Each of the color branches will have six number branches. The order of the branches does not affect the number of possible outcomes. In this case, any order of branches will produce 48 outcomes.

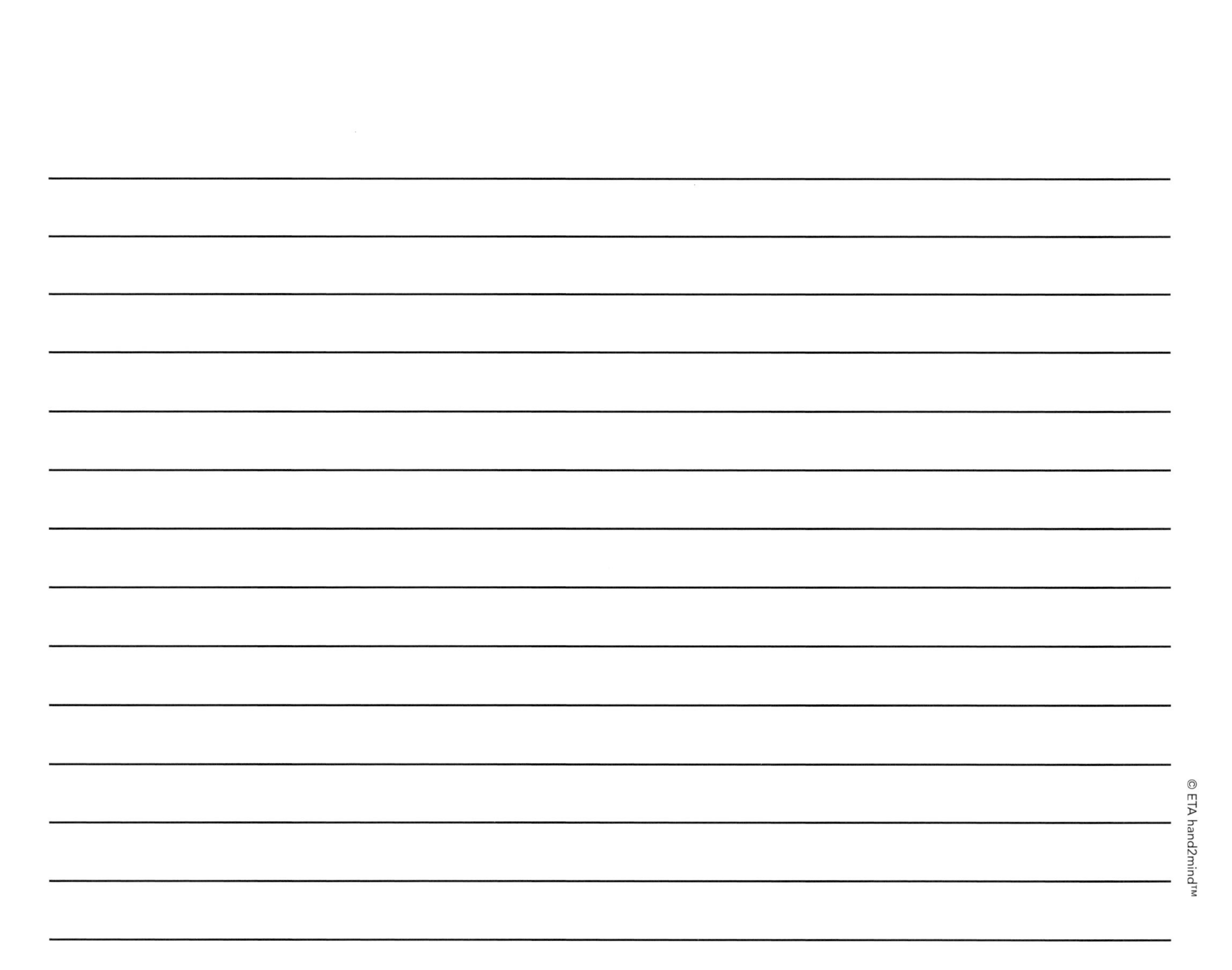

BLM 1

Centimeter Grid

Name ____________________

Name ______________________________

BLM 3

1-cm Number Lines

Name ______________________________

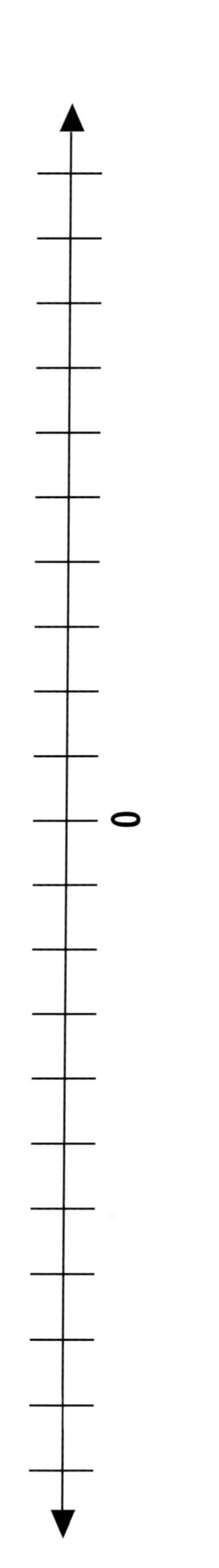

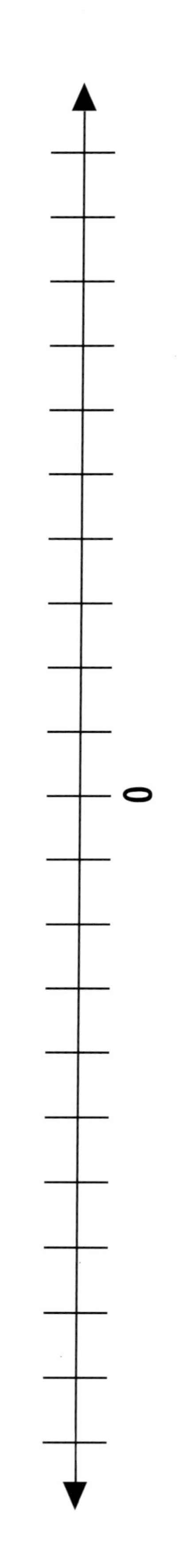

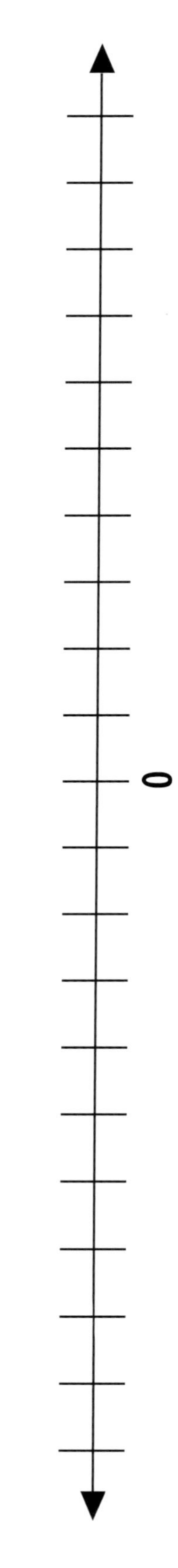

Name ______________________________

BLM 4

Algeblocks® Basic Mat

+

−

Name ____

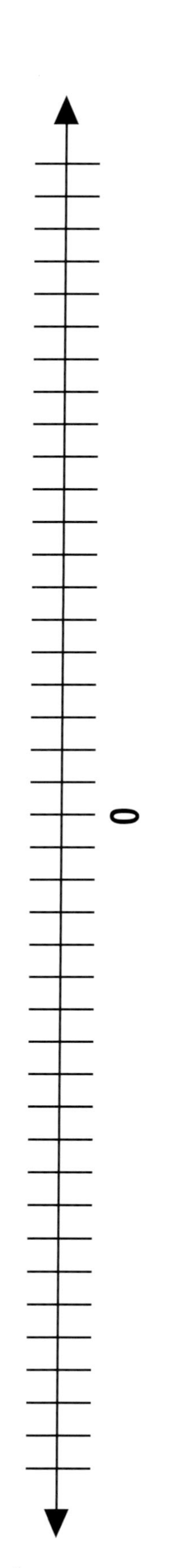

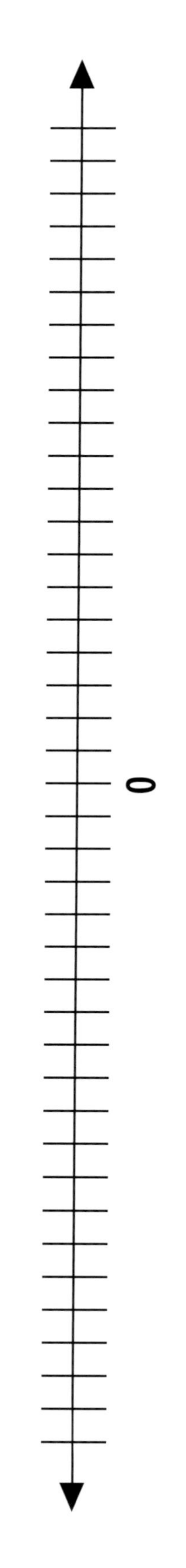

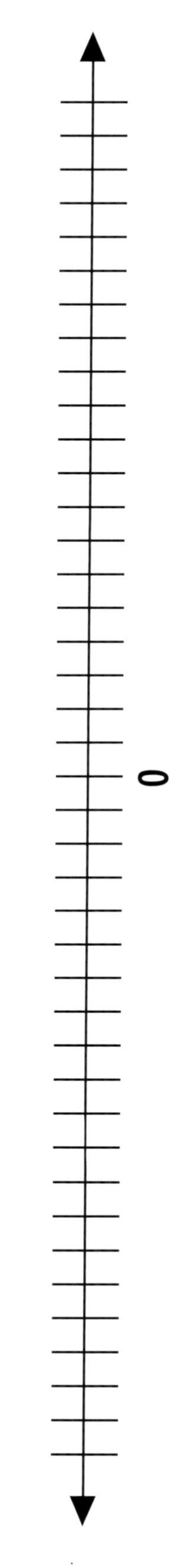

Name ______________________________

+	I	I
I		+
I	+	+

BLM 6

Algeblocks® Quadrant Mat

BLM 7

Algeblocks® Sentences Mat

Name ______________________________

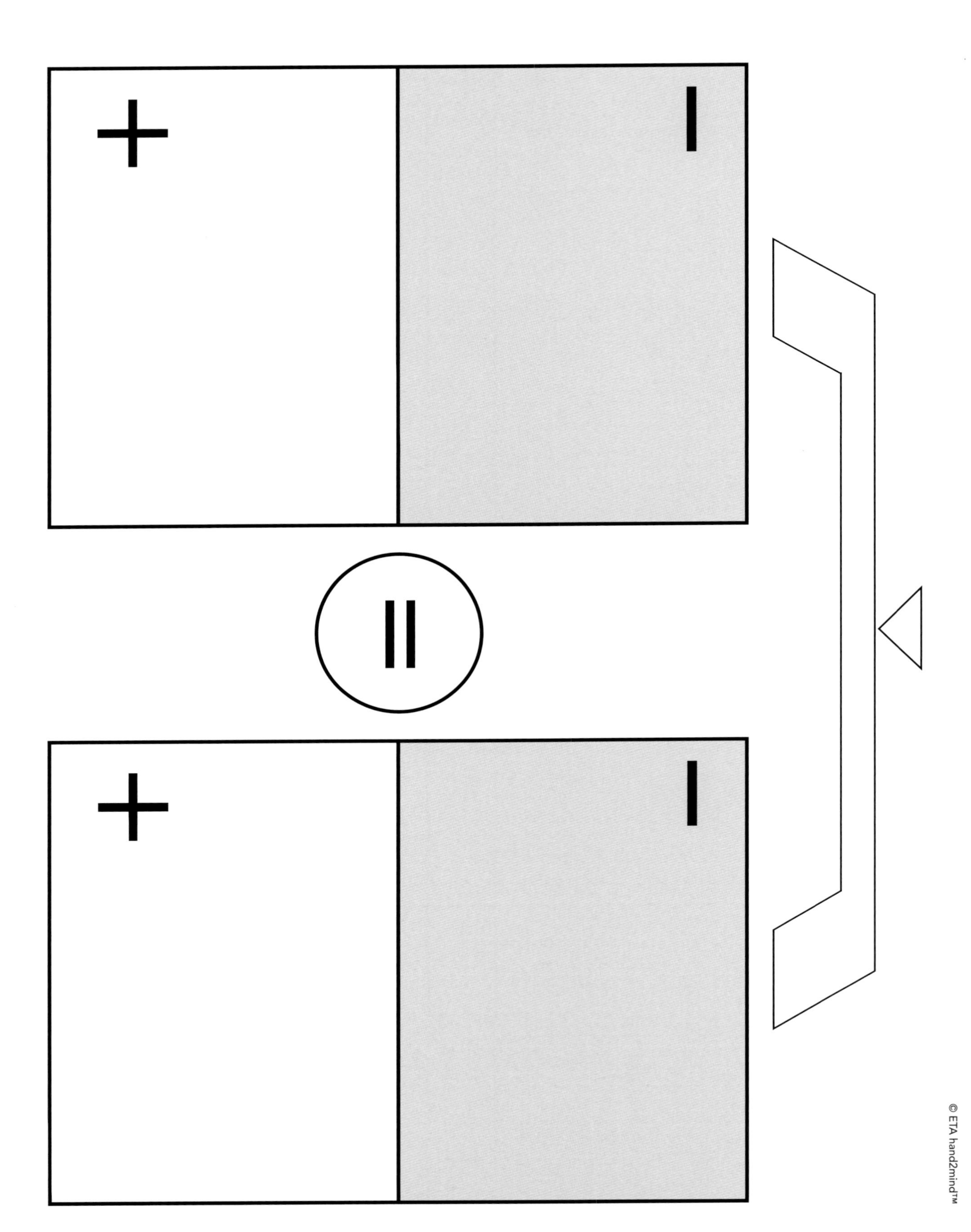

Name ______________________________

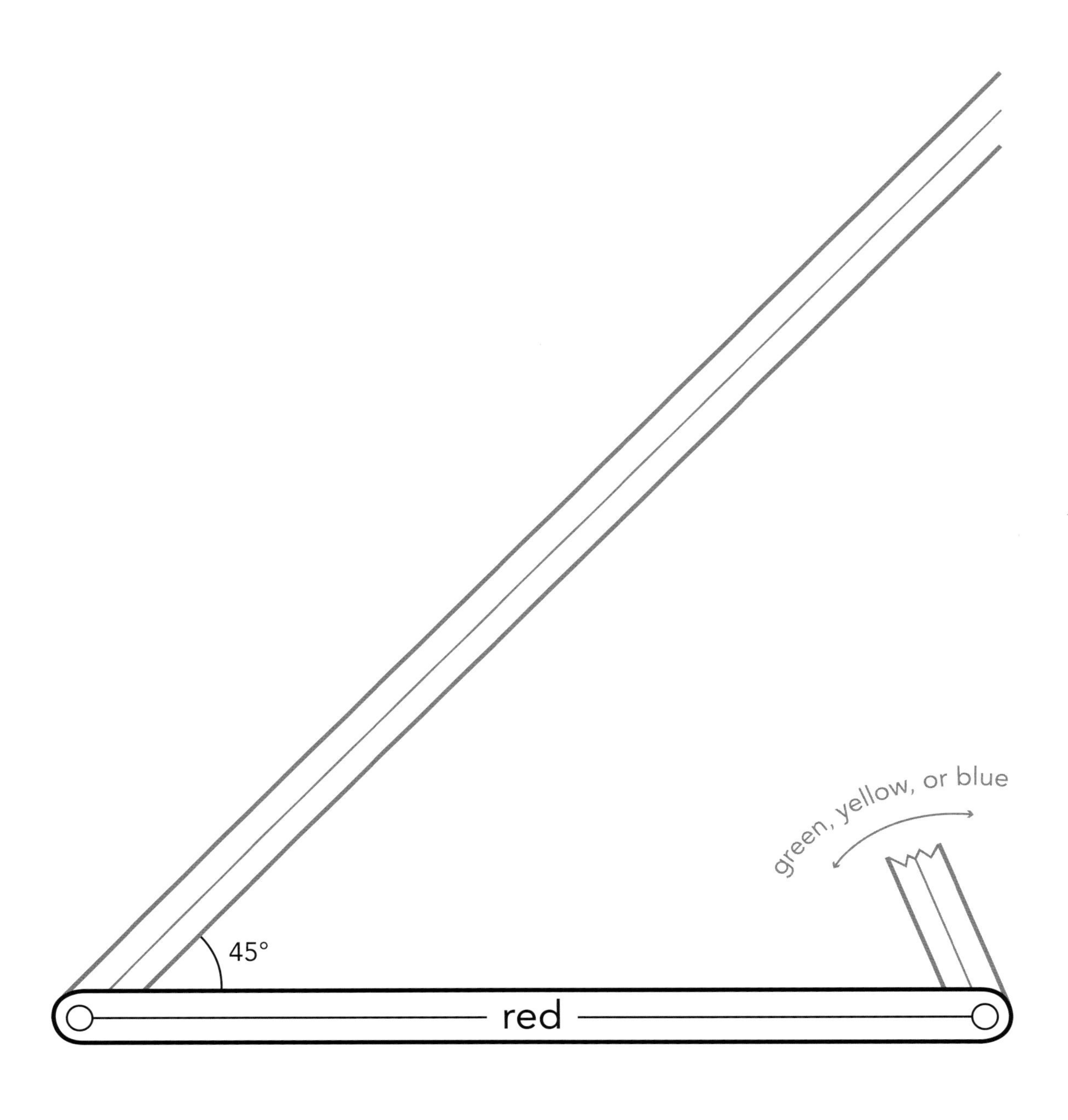

BLM 9

4-Column Recording Chart

Name ______________________________

Name ______________________________

BLM 10

Centimeter Dot Paper

BLM 11

1-Inch Triangular Grid Paper

Name ______________________________

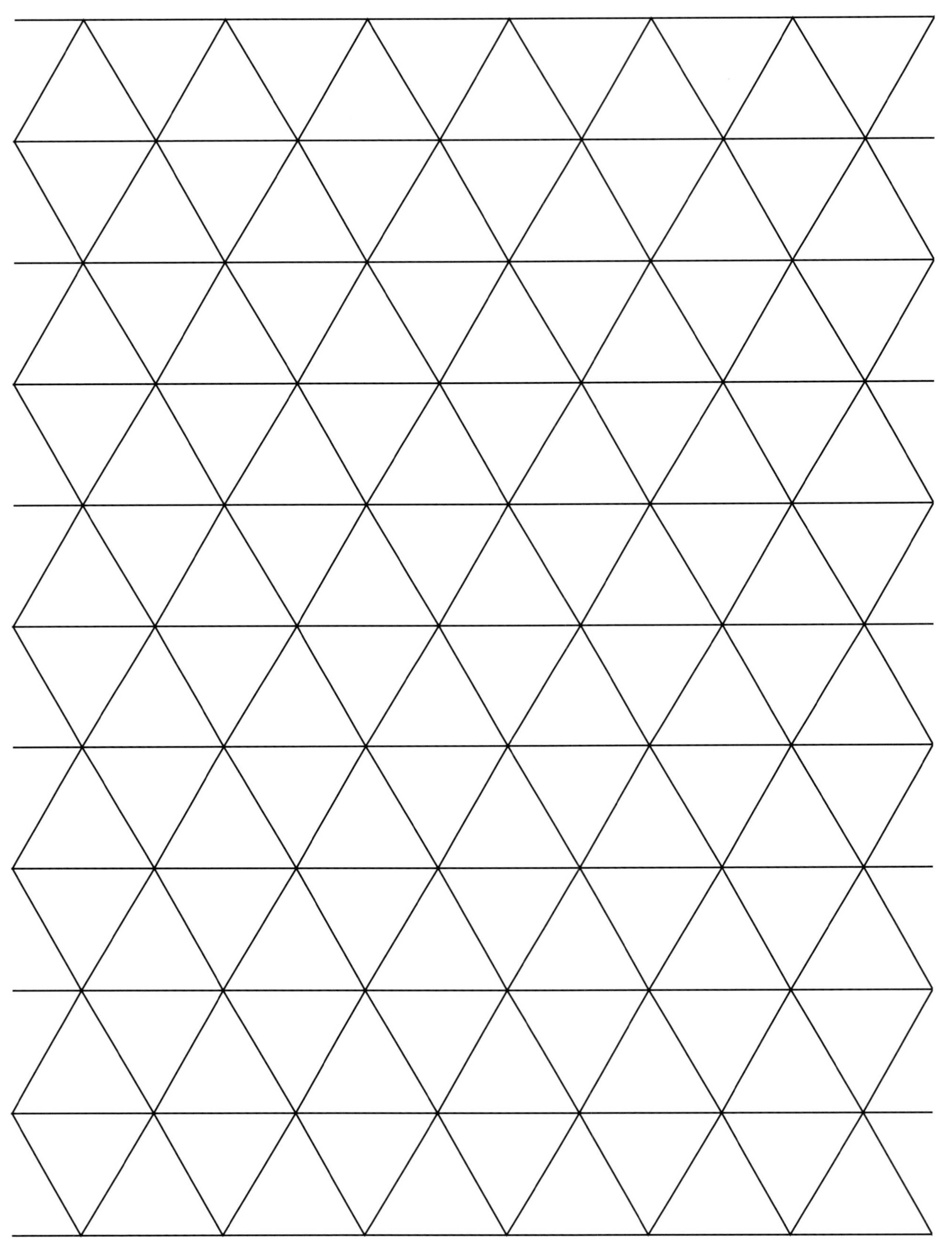

Name ______________________________

BLM 12

30-Trial Recording Chart

Trial	First Cube	Second Cube

BLM 13

10 x 10 Grid

Name ____

Glossary of Manipulatives

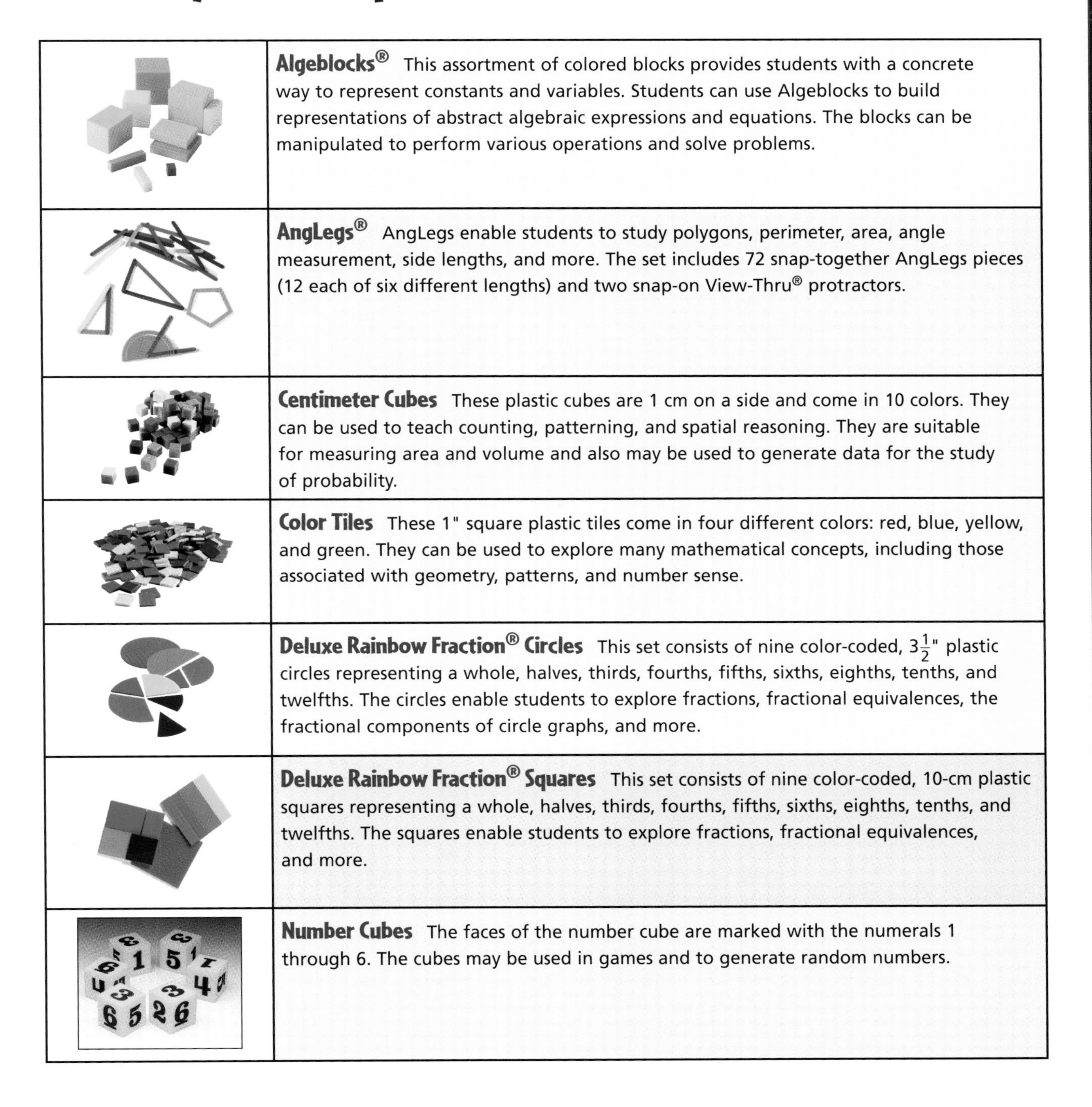

	Algeblocks® This assortment of colored blocks provides students with a concrete way to represent constants and variables. Students can use Algeblocks to build representations of abstract algebraic expressions and equations. The blocks can be manipulated to perform various operations and solve problems.
	AngLegs® AngLegs enable students to study polygons, perimeter, area, angle measurement, side lengths, and more. The set includes 72 snap-together AngLegs pieces (12 each of six different lengths) and two snap-on View-Thru® protractors.
	Centimeter Cubes These plastic cubes are 1 cm on a side and come in 10 colors. They can be used to teach counting, patterning, and spatial reasoning. They are suitable for measuring area and volume and also may be used to generate data for the study of probability.
	Color Tiles These 1" square plastic tiles come in four different colors: red, blue, yellow, and green. They can be used to explore many mathematical concepts, including those associated with geometry, patterns, and number sense.
	Deluxe Rainbow Fraction® Circles This set consists of nine color-coded, $3\frac{1}{2}$" plastic circles representing a whole, halves, thirds, fourths, fifths, sixths, eighths, tenths, and twelfths. The circles enable students to explore fractions, fractional equivalences, the fractional components of circle graphs, and more.
	Deluxe Rainbow Fraction® Squares This set consists of nine color-coded, 10-cm plastic squares representing a whole, halves, thirds, fourths, fifths, sixths, eighths, tenths, and twelfths. The squares enable students to explore fractions, fractional equivalences, and more.
	Number Cubes The faces of the number cube are marked with the numerals 1 through 6. The cubes may be used in games and to generate random numbers.

	**Pattern Blocks** Pattern Blocks come in six different color-shape varieties: yellow hexagons, red trapezoids, orange squares, green triangles, blue parallelograms (rhombuses), and tan rhombuses. They can be used to teach concepts from all strands of mathematics; for example, algebraic concepts such as patterning and sorting, as well as geometry and measurement concepts such as transformations, symmetry, and area. The blocks can also be used to study number and fraction relationships.
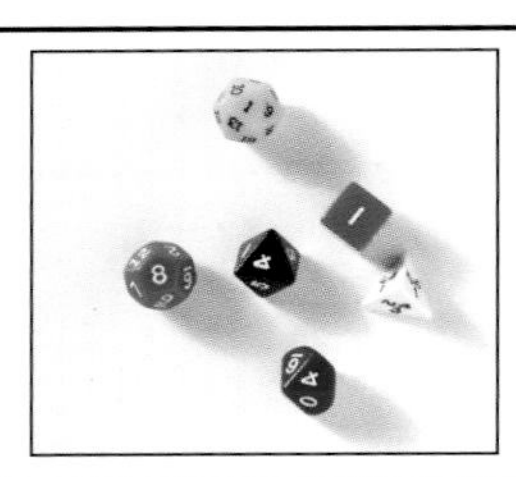	**Polyhedral Dice Set** This set consists of 4-, 6-, 8-, 10-, 12-, and 20-sided dice that may be used for a variety of probability activities. Dice may be used to generate data for number and operations activities as well as for data analysis.
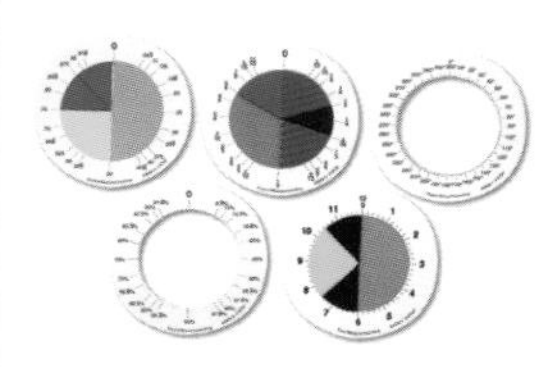	**Rainbow Fraction® Circle Rings** Each of these five plastic rings fits around the Deluxe Rainbow Fraction Circles, allowing the various sectors to be measured. This set consists of a Degree Measurement Ring, a Fraction Measurement Ring, a Decimal Measurement Ring, a Percent Measurement Ring, and a Time Measurement Ring.
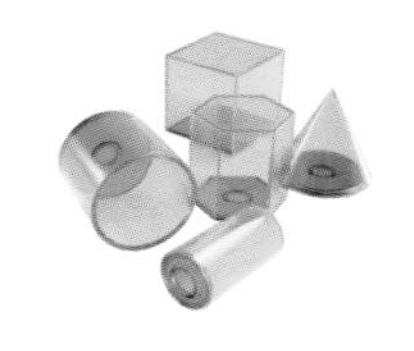	**Relational GeoSolids®** Relational GeoSolids is a set of 14 three-dimensional shapes that can be used to teach about prisms, pyramids, spheres, cylinders, cones, and hemispheres. GeoSolids facilitate classroom demonstrations and experimentation. The shapes can be filled with water, sand, rice, or other materials to give students a concrete framework for the study of volume.
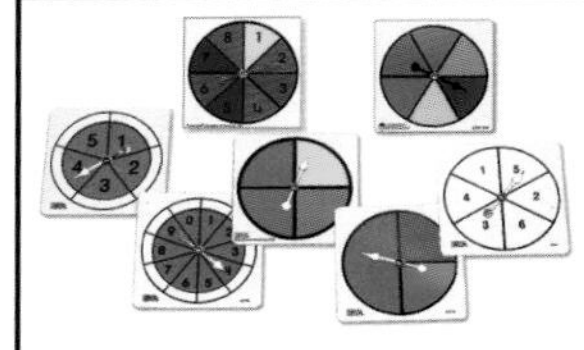	**Spinners** Spinners enable students to study probability and to generate numbers and data lists for number operations and data analysis.
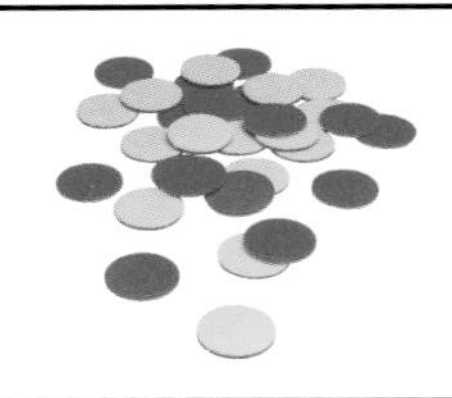	**Two-Color Counters** These versatile counters are thicker than most other counters and easy for students to manipulate. They can be used to teach number and operations concepts such as patterning, addition and subtraction, and multiplication and division. Counters also can be used to introduce students to basic ideas of probability.
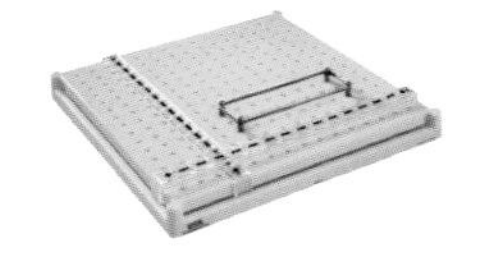	**XY Coordinate Pegboard** The XY Coordinate Pegboard can be used to graph coordinates in one, two, or four quadrants; to show translations of geometric figures; to display data in various forms; and to demonstrate numerous algebraic concepts and relationships.

Index

Boldface page numbers indicate when a manipulative is used in the Try It! activity.